Hybrid Modelling and Multi-Parametric Control of Bioprocesses

Special Issue Editor
Christoph Herwig

MDPI • Basel • Beijing • Wuhan • Barcelona • Belgrade

MDPI

Special Issue Editor
Christoph Herwig
Vienna University of Technology
Austria

Editorial Office
MDPI AG
St. Alban-Anlage 66
Basel, Switzerland

This edition is a reprint of the Special Issue published online in the open access journal *Bioengineer-ing* (ISSN 2306-5354) from 2016–2018 (available at: http://www.mdpi.com/journal/bioengineering/special issues/hybrid-modelling).

For citation purposes, cite each article independently as indicated on the article page online and as indicated below:

Lastname, F.M.; Lastname, F.M. Article title. *Journal Name*. **Year**. *Article number*, page range.

First Edition 2018

Image courtesy of Christoph Herwig

ISBN 978-3-03842-745-2 (Pbk)
ISBN 978-3-03842-746-9 (PDF)

Table of Contents

About the Special Issue Editor

Christoph Herwig, Prof. Dr. techn. Dipl.-Ing., bioprocess engineer from RWTH Aachen, worked in industry in the design and commissioning of large chemical facilities prior to entering his interdisciplinary PhD studies at EPFL, Switzerland, in bioprocess identification. Subsequently, he positioned himself at the interface between bioprocess development and facility design of biopharmaceutical fa-cilities. Since 2008, he has been full professor for biochemical engineering at the Vienna University of Technology. His research area focuses on the development of methods for integrated, science-based and efficient bioprocess development along PAT and QbD principles. The product fields he is inter-ested in are circular economy and biopharmaceuticals within industry driven projects.

Preface to "Hybrid Modelling and Multi-Parametric Control of Bioprocesses"

The goal of bioprocessing is to optimize process variables, such as product quantity and quality, in a reproducible, scalable, and transferable manner. However, bioprocesses are highly complex. A large number of process parameters and raw material attributes exist, which are highly interactive, and may vary from batch to batch. Those interactions need to be understood, and the source of variance must be identified and controlled.

While purely data-driven correlations, such as chemometric models of spectroscopic data, may be employed for the understanding of how process parameters are related to process variables, they can hardly be deployed outside of the calibration space. Currently, mechanistic models, models based on mechanistic links and first principles, are in the focus of development. They are perceived to allow transferability and scalability, because mechanistics can be extrapolated. Moreover, the models deliver a large range of hardly measureable states and physiological parameters.

For implementation of mechanistic models, however, models need to be simplified and linked to process parameters for real time execution. For this, hybrid models, and hence links between data driven and mechanistic models, may be helpful solutions.

Moreover, models need to be deployed in the control context: Bioprocesses need to be controlled, on the one hand, on different parameters simultaneously (e.g., constant precursor concentration and specific growth rate) and, on the other hand, may have different objective functions (maximum productivity and correct product quality). Hence, novel solutions and case studies for multiple input and output controls need to be developed, as they already exist in other market segments.

The current Special Issue aims to display current solutions and case studies of development and deployment of hybrid models and multi-parametric control of bioprocesses. It includes:

- Models for Bioprocess Monitoring
- Model for Bioreactor Design and Scale Up
- Hybrid model solutions, combinations of data driven and mechanistic models.
- Model to unravel mechanistic physiological regulations
- Implementation of hybrid models in the real-time context
- Data science driven model for process validation and product life cycle management.

Christoph Herwig
Special Issue Editor

bioengineering

MDPI

Article

Fluorometric In Situ Monitoring of an *Escherichia coli* Cell Factory with Cytosolic Expression of Human Glycosyltransferase GalNAcT2: Prospects and Limitations

Karen Schwab *,†, Jennifer Lauber † and Friedemann Hesse

Biberach University of Applied Sciences, Institute of Applied Biotechnology (IAB), 88400 Biberach, Germany; Lauber.Jennifer@web.de (J.L.); hesse@hochschule-bc.de (F.H.)
* Correspondence: schwab@hochschule-bc.de or karen.schwab@gmail.com; Tel.: +49-0-7351-582-442
† These authors contributed equally to this work.

Academic Editor: Christoph Herwig
Received: 16 October 2016; Accepted: 16 November 2016; Published: 21 November 2016

Abstract: The glycosyltransferase HisDapGalNAcT2 is the key protein of the *Escherichia coli* (*E. coli*) SHuffle® T7 cell factory which was genetically engineered to allow glycosylation of a protein substrate in vivo. The specific activity of the glycosyltransferase requires time-intensive analytics, but is a critical process parameter. Therefore, it has to be monitored closely. This study evaluates fluorometric in situ monitoring as option to access this critical process parameter during complex *E. coli* fermentations. Partial least square regression (PLS) models were built based on the fluorometric data recorded during the EnPresso® B fermentations. Capable models for the prediction of glucose and acetate concentrations were built for these fermentations with rout mean squared errors for prediction (RMSEP) of 0.19 $g \cdot L^{-1}$ and 0.08 $g \cdot L^{-1}$, as well as for the prediction of the optical density (RMSEP 0.24). In situ monitoring of soluble enzyme to cell dry weight ratios (RMSEP 5.5×10^{-4} µg *w/w*) and specific activity of the glycosyltransferase (RMSEP 33.5 $pmol \cdot min^{-1} \cdot µg^{-1}$) proved to be challenging, since HisDapGalNAcT2 had to be extracted from the cells and purified. However, fluorescence spectroscopy, in combination with PLS modeling, proved to be feasible for in situ monitoring of complex expression systems.

Keywords: *E. coli* SHuffle® T7; Glycosyltransferase GalNAcT2; In Situ monitoring; soft sensor; fluorescence spectroscopy

1. Introduction

The successful expression of target proteins that require post-translational modifications, such as glycosylations or disulfide bond formation, remains a challenge in *Escherichia coli*. These modifications are indispensable for protein folding, stability, and activity. Disulfide bond formation is usually compartmentalized in the periplasm of *E. coli*. The oxidizing environment and the presence of various chaperones in the periplasm enable the oxidation of sulfhydryl groups between two cysteine side chains resulting in a covalent disulfide bond [1,2]. Nevertheless, *E. coli* periplasm is poorly adapted for the production of multi-disulfide-bonded proteins in high yields since the periplasmic space is significantly smaller than the cytoplasmic space [3]. Therefore, another option is the expression of recombinant protein in the cytoplasm, whereby the chance for inclusion body formation is given. High expression rates and the lack of chaperones mediating correct folding and disulfide bond formation supports the accumulation of insoluble protein. Hence, some strains were engineered to enable the formation of disulfide bonds in the cytoplasm. These strains carry mutations in both the thioredoxin reductase (*trxB*) and the glutathione reductase (*gor*) genes to provide a less reducing cytoplasmic environment [1,4,5]. In addition to the mutations in *trxB* and *gor*, SHuffle® T7 carries the chromosomally-integrated gene for the disulfide bond isomerase DsbC without the signal sequence [1]. To engineer an *E. coli* cell factory capable of glycosylating a protein substrate within the cytoplasm requires the expression of an additional enzyme. In the work presented here, the recently established *E. coli* cell factory derived from the SHuffle® T7 strain was used. The proposed strain by Lauber et al. [6] was genetically engineered to express a functional recombinant human-derived glycosyltransferase. It could be shown that the co-expression of two redox folding helpers enabled the formation of a soluble enzyme with four disulfide bonds. This contributed to the development of a glycosylation system in *E. coli* for the transfer of a GalNAc-residue to a protein substrate in the cytoplasm [6]. The selection of the media used for bacteria fermentations is not negligible and can have a major impact on the bacterial cell factory [7]. High initial glucose concentrations in batch cultivations rapidly deplete accompanied by acetate increases at the same time [7]. Jain et al. showed in their experiments, that the glucose concentrations regulated the growth rates in *E. coli* fed-batch cultivations. Hoffmann et al. [8] decreased inclusion body formation in *E. coli* fed-batch cultures expressing β-galactosidase-HIVgp41 fusion protein by maintaining low glucose concentrations. Whereas Luchner et al. [9] controlled the expression rate of soluble human superoxide dismutase in *E. coli* by limiting the induction. The group showed that slowing down the protein expression shifts the ratio of soluble protein to inclusion bodies towards the soluble product. Moreover, Hortsch and Weuster-Botz [7] showed that the enzymatic glucose release of the EnPresso® B medium can help to increase the expression of soluble formate dehydrogenase. They concluded that the consistently low glucose concentrations prevented the *E. coli* cells from metabolic overflow. Hence, the growth rates were reduced and the culture showed no acetate shoot up in combination with a pH drop.

Various process analyzers were already used over the last 15 years to set up non-invasive online monitoring systems for process parameters. Chemometric modeling was used to access the complex data structure generated with the used process analyzers [10–12]. Among the used methods, 2D fluorescence spectroscopy (2DFS) has been proven to be a highly valuable, sensitive, and reliable process analyzer which can be used for the prediction of substrate, product, and metabolite concentrations [9,13–15]. In general, principal components analysis (PCA) can be used to analyze the structure of the generated datasets, investigate multi-factorial relationships, and extract the relevant information [16,17]. Furthermore, partial least square regression (PLSR) can be applied to correlate offline measured process parameters, such as optical density or glucose concentrations (*y*-data), and the online recorded fluorescence scans (*x*-data) via linear regression [18]. Fluorescent components that are involved in cell growth and metabolism [19] change characteristically in the course of cultivation. The overall fluorescence signal of the bacterial culture is a mixture of fluorescence signals, which originate from components such as aromatic amino acids, ATP, NADH, FAD, and vitamins (riboflavin and pyridoxine). Furthermore, some fluorophores are also supplemented to the culture with the feed medium [14]. Cell growth and biomass formation are the most common monitored bioprocess parameters based on 2DFS. This was not only applied for *E. coli* cultivation processes [20], but also for *Saccharomyces cerevisiae* [21], *Pichia pastoris* [22], *Aspergillus niger* spores [23] and *Klebsiella pneumoniae* [24]. In addition, existing literature also shows that substrate, metabolite, and product concentrations [25,26] can be monitored in microbial cultivations but also in mammalian cultures [14,15]. Luchner et al. were even able to predict the concentration of soluble recombinant superoxide dismutase in *E. coli* fed-batch cultivations based on 2DFS [9]. The general attempt of all of these studies was to avoid complex and time consuming offline analyses, improve process understanding, and enable better process control.

The *E. coli* cell factory expressing soluble human-derived glycosyltransferase was chosen as the model system for this study. The cell factory expresses two helper proteins in order to support the formation of active HisDapGalNAcT2. Hence, the amount of functionally active enzyme in the cytoplasm was an indication of the cell factories performance. These process parameters were not accessible during the fermentation and had to be determined afterwards. The aim of this study was to evaluate if 2DFS based soft sensors can be applied for in situ prediction of difficult-to-access process parameters that require time-intensive and costly analytics. Furthermore, standard process parameters were monitored since a highly fluorescent medium with enzymatic glucose release was used, which was assumed to complicate the model generation. These models were built to illustrate the differences and challenges we were facing while calibrating PLSR models for the prediction of soluble protein to dry weight ratios and the specific activity of the glycosyltransferase. The study illustrated that it can be beneficial to use process analyzers to monitor all critical process parameters in real-time.

2. Material and Methods

2.1. Strain

All experiments were performed with *E. coli* SHuffle® T7 (C3026H, New England Biolabs, Frankfurt am Main, Germany) expressing the recombinant human glycosyltransferase fusion protein HisDapGalNAcT2 and the chaperones Erv1p and PDI [6]. The glycosyltransferase HisDapGalNAcT2, which was encoded on the plasmid pET23d(+)::*HisDapGalNAcT2*, was under the control of the T7-promotor and construction of the plasmid was described previously [6]. The two chaperone genes on the plasmid pMJS9 were under the control of the arabinose promotor. Plasmid pMJS9 [27] was kindly provided by L. W. Ruddock. Unless otherwise stated, chemicals and reagents were obtained from Sigma-Aldrich (Taufkirchen, Germany) or Roth (Carl Roth GmbH and Co. KG, Karlsruhe, Germany).

2.2. Pre-Cultivation

The *E. coli* pre-cultures to inoculate the EnPresso® B (BioSilta Oy, Oulu, Finland) batch process were cultivated in LB medium (120 µg·mL^{-1} ampicillin, 34 µg·mL^{-1} chloramphenicol and 0.2% glucose). The pre-cultures were inoculated with eight ceramic cryo beads containing the *E. coli* SHuffle® T7 strain. The pre-cultures were grown in 50 mL Falcon™ tubes (Fisher Scientific, Schwerte, Germany) containing 15 mL medium for 8 h at 37 °C/175 rpm.

2.3. Bioreactor Culture

A 2 L benchtop bioreactor BIOSTAT® Bplus (Sartorius, Göttingen, Germany) equipped with two rushton impellers was used for the cultivation with a working volume of 1.4 L. Temperature, pH, and aeration were set to 30 °C, pH 7, and 0.05 vvm, respectively. The oxygen saturation was kept constant at 60% via agitation starting with a lower limit of 100 rpm. The pH was controlled with 1M NaOH and 1M H$_2$SO$_4$. Furthermore 5% (*v/v*) DOW CORNING® medical antifoam (Dow Corning, Midland, MI, USA) was used.

2.4. Batch Cultivation with EnPresso®B Medium

The complex predefined EnPresso® B medium including booster tablets was used as growth medium. The EnPresso® B tablets were dissolved in sterile demineralized water and the solubilized medium was transferred into the sterilized bioreactor. Following the protocol provided by the manufacturer, the pre-culture was used for inoculation and a final OD$_{600}$ of ≤0.04 was measured in the bioreactor after inoculation. All fermentations were supplemented with 120 µg·mL^{-1} ampicillin, 34 µg·mL^{-1} chloramphenicol, and the amylase for glucose release. After 15 h cultivation, booster tablets and amylase were added according to the manufacturer's protocol. The pre-induction of the pMJS9 encoded gene products was carried out in the presence of 0.5% *w/v* arabinose added to the bioreactor 30 min after the booster tablets. Isopropyl-β-D-thiogalactopyranosid (IPTG) was added after another 30 min to a final concentration of 1 mmol·L^{-1} to induce expression of the glycosyltransferase HisDapGalNAcT2 [6]. Over the following 23.5 h 12 samples were taken from the bioreactor for offline analysis starting with the first sample after addition of arabinose. The samples were stored on ice during the cultivation.

2.5. Offline Analytics

OD_{600} was measured using the photospectrometer Ultrospec 3100 pro (Amersham Bioscience Europe, Freiburg, Germany). Acetate and glucose concentrations were determined enzymatically with the Konelab Arena XT (Thermo Scientific, Waltham, MA, USA) using an acetate kit (R-Biopharm AG, Darmstadt, Germany) and a glucose kit (Thermo Fisher Scientific, Waltham, MA, USA). Bacterial dry matter was determined by centrifugation of 5 ml cell suspension. The pellet was re-suspended in PBS for the transfer to a pre-weighed test tube. This was followed by an additional centrifugation step. The supernatant was discarded and the bacteria pellets in the test tubes were dried at 105 °C for 24 h and re-weighed.

2.6. Purification of Soluble Human Glycosyltransferase

A cell pellet derived from a 5 mL culture fraction was re-suspended in 630 µL extraction buffer (50 mmol·L^{-1} Tris, 300 mmol·L^{-1} NaCl, pH 8) containing 70 µL lysozyme, 1.5 µL DNAse I and 25 µL protease inhibitor (complete protease inhibitor cocktail tablet, F. Hoffmann La-Roche AG, Switzerland). The bacterial suspension was cooled on ice for 30 min and sonicated for 3 min on ice. The cell lysate was centrifuged at 4 °C, 16100× g for 10 min, and the supernatant was passed through a 0.45 µm filter (Merck Millipore, Darmstadt, Germany). The polyhistidine-tagged protein HisDapGalNAcT2 was purified using Ni-NTA spin columns (Qiagen, Hilden, Germany) with washing buffer (50 mmol·L^{-1} Tris, 300 mmol·L^{-1} NaCl, 20 mmol·L^{-1} imidazole) and elution buffer (50 mmol·L^{-1} Tris, 300 mmol·L^{-1} NaCl, 500 mmol·L^{-1} imidazole) adjusted to pH 8 prior to use [6]. The protein concentration in the eluent fraction was determined by employing a BCA-assay.

2.7. Human Glycosyltransferase Activity Assay

The activity of HisDapGalNAcT2 was determined using a glycosyltransferase activity kit (EA001, R & D Systems Europe Ltd., Abingdon, UK) as described previously [6]. The activity of each sample was determined using 0.5 µg soluble protein.

2.8. Online Data Collection

The multi-wavelength excitation/emission matrices (EEM) were recorded with a BioView® system (Delta, Hørsholm, Denmark) equipped with a fiber optic assembly especially developed to fit into a 19 mm port. The benchtop bioreactor was equipped with a 20 cm stainless steel casing and the fluorescence sensor was inserted after autoclaving. A full EEM consists of 120 wavelength pairs with an excitation range from 270 to 550 nm and emissions recorded from 290 to 590 nm with a 20 nm interval. The scans were vectorized into 2-way arrays and used as x-data for the chemometric modeling. The gain of the fluorescence spectrometer was set to 1100 and the EEMs were recorded with a measurement interval of 5 min during fermentation. Only EEMs taken after booster addition were used for the chemometric modeling.

2.9. Chemometric Modeling

MATLAB version 8.4.0 (MathWorks, Natick, MA, USA) in combination with the PLS-toolbox version 7.9.5 (Eigenvector Research Inc., Manson, WA, USA) was used for chemometric modeling [28]. A detailed description is provided elsewhere in the literature [17,18,29]. The EEM data were preprocessed by background subtraction to the first scan after booster addition. Offline measured values and corresponding scans were used for the calibration of PLS regression models for the prediction of OD_{600}, acetate, and glucose concentration. For the PLS model, regarding the ratio of soluble protein to dry cell weight, only samples 21 h after inoculation were used as input data for the calibration model. For the PLS model concerning the specific activity of HisDapGalNAcT2, the offline data was complemented using a double Boltzmann fitting operated in Origin 9.1G (OriginLab Inc., Northampton, MA, USA). Based on the fit, the resulting y-data in 30 min intervals and corresponding

EEM were applied for the correlation starting with all scans recorded between 24.5 h after inoculation and harvest. The SIMPLS algorithm was used for PLSR model calibration in combination with venetian blinds as method for cross-validation applying six splits and one sample per split for all models. Three of the available datasets (run I–III) were used for calibration and cross-validation. Calculated RMSE and RMSECV were used to assess the performance of the model. All EEMs recorded after booster addition were fed to the respective PLS model for prediction. In order to evaluate the robustness of the calibration model, x-data recorded during an additional fermentation (run IV) was only predicted using the selected model. The predicted response variables of all fermentations were compared to the offline-determined values.

3. Results and Discussion

The expression of the human-derived, soluble, and functional glycosyltransferase HisDapGalNAcT2 represents the key factor in establishing this particular *E. coli* cell factory. The final purpose of this cell factory will be to enable the transfer of a GalNAc-residue to a protein substrate in vivo. The expression strategy for HisDapGalNAcT2 did follow a temporal sequence. Two chaperones (sulfhydryl oxidase Erv1p and protein disulfide isomerase PDI [6,27]) were induced via arabinose and mediated folding and disulfide bond formation of the HisDapGalNAcT2. The glycosyltransferase was induced by IPTG 30 min after the two redox folding helpers.

3.1. EnPresso® B Batch Cultivations and Chemometric Modeling of Process Parameters

Four batch fermentations were carried out using EnPresso® B medium and the *E. coli* SHuffle® T7 strain expressing the recombinant glycosyltransferase HisDapGalNAcT2. EnPresso® B medium consists of three main components: (1) medium tablets, (2) booster tablet and (3) amylase for controlled glucose release. The components of the tablets are not stated by the manufacturer but it is known that a polysaccharide is an ingredient of the booster and the medium tablets. Following the protocol provided by the manufacturer, the booster was added to the medium 15 h after inoculation and 1 h before the chaperones were induced. A sufficient nutrition of the cell factory was achieved by addition of amylase at the inoculation and again together with the booster [7]. The measured glucose concentrations ranged between 0.3–2.3 g·L^{-1} during the cultivations. Low cell growth with doubling times of 172 ± 4.4 min was observed during the first 15 h after inoculation prior to booster addition. The booster addition accelerated the cell growth and doubling times of 110 ± 17.8 min were observed. The growth rates declined already 2 h after booster addition and doubling times between 30–60 h within the following 20 h were observed. Nevertheless, all EnPresso® B cultivations reached higher OD$_{600}$ values at the end of the process in comparison to the cultivation with LB-medium (Figure S1).

Offline values and corresponding EEMs of three fermentations (runs I–III) were used for calibration and cross-validation of the respective soft sensors. Resulting PLSR models were selected and evaluated based on preferably low rout mean squared errors for calibration and prediction (RMSEC and RMSEP) in combination with $R^2 > 0.9$ for calibration (R^2_{cal}) and validation (R^2_{CV}) if possible (Table 1). The number of latent variables (LVs) required for each model was minimized and it was aimed for a maximum of captured x- and y-variance simultaneously. The PLSR models were, furthermore, applied to predict the respective y-values during an additional batch cultivation using EnPresso® B (run IV). This was done to investigate the robustness and predictive power of these models.

3.2. Overall Batch Behavior Evaluated by Principal Component Analysis

Principal component analysis (PCA) was used to investigate the structure of the fluorometric datasets and enabled the identification of differences between the batch cultivations with EnPresso® B medium prior to PLS modeling. The relation of the individual EEMs to each other can be displayed in the PCA scores plot. EEMs with similar scores are considered similar. The score values calculated for the four fermentations formed similar trajectories on the PCA score plot (Figure 1) which were compared and put into relation. The fluorometric dataset was not preprocessed, but the scores plot showed that 99.8% of the variance in the dataset was already captured by two principal components (PC).

The similar score values of batch cultivation I and II indicated that they had the same background fluorescence. The trajectory of batch cultivation III differed mainly on PC1 from cultivation I and II. Furthermore the trajectory of cultivation IV differed on PC1 and PC2 from all other cultivations. It was suspected that lot to lot variability of the EnPresso® B medium or the amylase performance might have caused this variability in the datasets. This will be discussed in the following chapters.

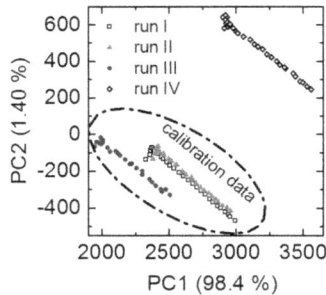

Figure 1. PCA scores plot. The EEM raw data was used as input for the PCA. The PCA scores plot shows the differences between the four batch cultivations with EnPresso® B medium. For clarity only every tenth point is shown. The circle indicates the cultivations used for PLS model calibration and run IV was used as test dataset to predict the target parameters. The cultivation run IV was carried out with six months time lag to the other cultivations.

3.3. Prediction of Acetate Concentrations and Optical Density

The calculated PLS models for the prediction of acetate concentrations and OD_{600} values showed in both cases a calculated $R^2_{cal} > 0.98$ (Table 1). The quality of the fit was also evaluated with the help of the predicted versus measured plots (Figure 2A,B). All OD_{600} values and acetate concentrations were located closely to the target line. Captured variances of 99.9% for the x-data and 99.85% for the y-data with only four LVs were achieved for the PLS model which was built to predict acetate concentrations. In addition, 81.87% of the variance in the x-data and 89.81% of the variance in the y-data was captured for the OD_{600} model also with four LVs (Table 1). The correlation of predicted values and offline measured acetate concentrations and OD_{600} values was good for all batch cultivations used for model generation (Figure 3A,B). Occasionally samples taken during batch cultivation run IV were analyzed offline. The measured values did fit the prediction for acetate concentrations and OD_{600} (Figure 3A,B). This led to the assumption that the proposed soft sensors were reliable and allowed the online prediction of acetate concentrations and OD_{600}.

Table 1. Quality attributes of the PLS models for the prediction of glucose and acetate concentrations, OD_{600}, the ratio of soluble protein to cell dry matter, and the specific activity of HisDapGalNAcT2.

Target	LV	R^2_{cal}	R^2_{CV}	RMSEC	RMSEP
Glucose concentration $(g \cdot L^{-1})$	3	0.93	0.88	0.14	0.19
Acetate concentration $(g \cdot L^{-1})$	4	0.99	0.97	0.05	0.08
OD_{600}	4	0.99	0.97	0.02	0.24
Specific activity GalNAcT2 $(pmol \cdot min^{-1} \cdot \mu g^{-1})$	3	0.65	0.59	30.7	33.5
Ratio of soluble GalNAcT2 / Dry matter $(\mu g \ w/w)$	3	0.86	0.74	4×10^{-4}	5.5×10^{-4}

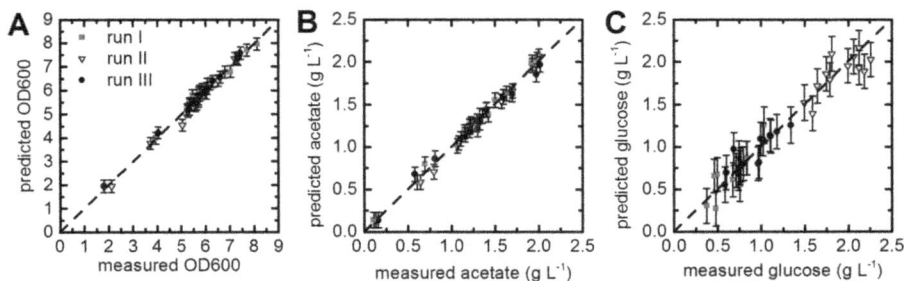

Figure 2. Prediction versus reference plots. PLS regression models were built for OD_{600} (**A**), acetate (**B**), and glucose concentrations (**C**). Online and offline data of three cultivations was used for calibration and cross-validation of the PLS models. Vectorized EEMs taken at the sample points (*x*-data) and the offline measured values (*y*-data) were correlated. Predicted versus reference plots of the PLS models did show good correlation.

3.4. Prediction of Substrate Concentrations

Establishing a soft sensor for the online prediction of glucose concentrations was challenging. Since the glucose supply of the culture was accomplished enzymatically through amylase (glucoamylase) [30]. The same batch of predefined EnPresso® B medium was used for all cultivations. The glucose concentrations determined for samples taken from the bioreactor during batch cultivation run I and run III ranged between 0.5 $g·L^{-1}$ and 1.5 $g·L^{-1}$. However, the glucose concentrations measured during batch cultivation run II were higher, ranging from 1.5 $g·L^{-1}$ up to 2.3 $g·L^{-1}$. Despite the differences between the three cultivations concerning the measured glucose concentrations, a PLS model was calibrated and cross-validated. The resulting model with three LVs and an R^2_{cal} of 0.94, R^2_{CV} of 0.88, and a RMSEP of 0.19 $g·L^{-1}$ was accepted (Table 1), since the predicted versus measured plot showed good correlation (Figure 2C). Glucose concentrations predicted for cultivations included in the model generation (run I–III) were in good accordance with the offline measured values (Figure 3C). The predicted glucose concentrations of batch fermentation run IV were roughly 1 $g·L^{-1}$ higher than the offline measured values (Figure 3C). Nevertheless, predicted and measured values showed the same trend over cultivation time. The discrepancies between the offline measured values and predicted glucose concentration might be explained as follows: Glucose is a non-fluorescent compound, but glucose uptake and consumption by the cells has an impact on the pattern of fluorescent components in the culture. Thus, chemometric models for glucose prediction are generally based on these patterns, so-called secondary effects. The EnPresso® B medium contained a polysaccharide and the glucose supply was regulated through amylase. The information about how much glucose was released over time was not available, because the glucose was continuously metabolized by the cell factory.

The observed circumstances let to the assumption that either the enzyme used in run IV or the cell metabolism behaved differently, which was already suspected based on the PCA results. First, a different enzyme lot was used for this particular cultivation. Second, the medium was stored as separately-wrapped tablets and, although the same medium lot was used, differences in the appearance of the tablets were observed due to storage. The largest difference concerning the color and the solubility in water was noticed between the medium tablets used for the first tree cultivations (runs I–III) and for cultivation IV. This validation run IV was conducted with a time lag of six months to the other cultivations.

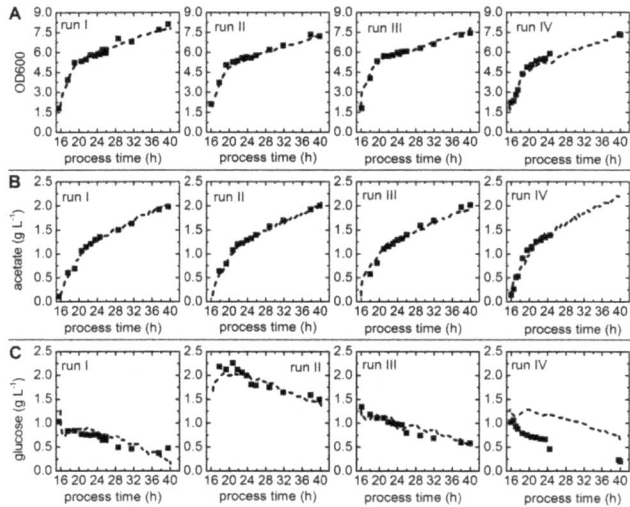

Figure 3. Correlation of predicted and offline measured process parameters. Predicted values (dashed) were compared to offline values (squares). Runs I–III were used for calibration and internal cross-validation of the soft sensors predicting OD_{600} (**A**), acetate (**B**), and glucose concentrations (**C**). Run IV was not included in model generation; this cultivation was used to test the selected model.

3.5. Chemometric Modeling of the Cell Factory's Efficiency

The proposed *E. coli* cell factory is a complex expression system with the objective to perform posttranslational changes to a protein substrate in vivo. Real-time monitoring of the glycosyltransferase specific activity during cultivation might be an advantage since the enzyme was not the final product of this process, but it was an indicator for the cell factory's efficiency. Since the ability to glycosylate a protein substrate is always directly related to the concentration of functional active HisDapGalNAcT2 in the cytoplasm.

3.6. Prediction of Soluble Protein to Biomass Ratio

A PLS regression model for the prediction of soluble protein accumulation in the cytoplasm of the cell factory was developed. Therefore, the ratio of captured soluble protein to dry cell matter was calculated and used as the *y*-value for the model calibration. It has to be considered that the offline measured soluble protein concentrations might have been biased due to the laborious purification method, since the chance to co-purify small amounts of host cell protein during the capture step with Ni-NTA spin columns was given. For this reason, only *y*-data determined for samples taken after 21 h cultivation time were used as input for the model. For these samples it was assumed that the amount of possibly co-purified host cell protein was negligible in comparison to the concentrations of the recombinant protein. The resulting calibration model with three LVs was able to capture 79.2% of the *x*-variance and 86.2% of the *y*-variance. The measured versus predicted plot of the PLS regression model showed a close correlation (Figure 4A) with correlation coefficients of $R^2_{cal} = 0.862$ and $R^2_{CV} = 0.744$ (Table 1). The scores plot showed that the EEMs of all four fermentations behaved similarly after preprocessing of the raw data by background subtraction (Figure 4B). This supported the supposition that the variations in the fluorescence data sets were related to the background fluorescence caused by the medium, like the PCA results, was already indicated. The PLS model was used to predict the ratio of soluble protein to cell dry weight based on the EEMs recorded during the cultivations. Figure 4C–E shows that offline and predicted values were in good accordance.

The soluble protein to cell dry weight ratio was also predicted for fermentation IV (Figure 4F). However, a corresponding offline dataset was not available for this run. Nevertheless, the predicted values showed the same progression over process time as observed for all other cultivations. This indicated that the selected PLS model interpreted the fluorometric dataset of this fermentation in the same way as all other fermentations. The ratio of soluble protein to cell dry matter of all cultivations did steadily increase to approximately 4×10^{-3} µg (*w/w*). However, a distinct increase of the ratio was observed for the first five hours after induction. This observation can be assigned to the second glucoamylase addition prior to induction, and consequently enhanced glucose release [7]. One hour after induction, the replication of the plasmids and the synthesis of the recombinant proteins most likely effected the cell growth. Diaz and Hernández [31] showed that cell metabolism and doubling times can be influenced by various parameters, such as plasmid size (pMJS9: 8.1 kBp and pET23d(+)::*HisDapGalNAcT2*: 5.4 kBp), copy number, over expression of homologous or heterologous genes, and their size. This particular *E. coli* cell factory was genetically engineered to express three heterologous genes: (1) sulfhydryloxidase (21.6 kDa); (2) protein disulfide isomerase (58.2 kDa); and (3) glycosyltransferase (61.7 kDa). This metabolic shift might have been crucial for the formation of active human recombinant HisDapGalNAcT2. In accordance with this, Luchner et al. showed, for human superoxide dismutase expressed in *E. coli*, that the ratio of active soluble protein to its aggregated inactive form was strongly dependant on the growth rate [9]. It was concluded that the uptake and processing of the HisDapGalNAcT2 by the chaperones, might only work when the cell growth and the glycosyltransferase expression was slowed down.

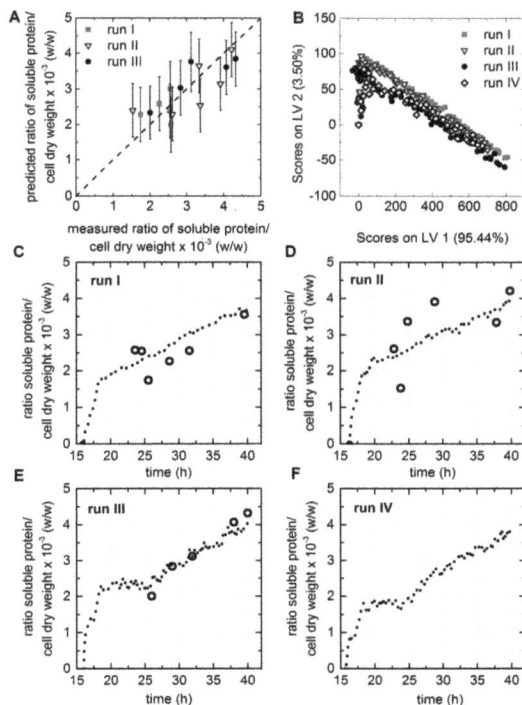

Figure 4. PLS model for the prediction of soluble protein to cell dry weight ratios. (**A**) The predicted versus reference plot of the PLS model with three LVs. (**B**) The scores values of all EEMs calculated for LV1 and LV2 are shown in the scores plot. (**C–F**) Offline (circles) and predicted values (dashed line) based on the selected PLS model are shown for cultivations I–IV.

3.7. Prediction of the Specific Activity of the HisDapGalNAcT2

The specific activity of the purified HisDapGalNAcT2 was determined following the activity assay described by Lauber et al. [6]. Three LVs were selected for the resulting PLS model predicting the specific activity of the HisDapGalNAcT2. The considerably low R^2_{cal} of 0.654 and R^2_{CV} of 0.591 did not meet the requirements described earlier, where correlation coefficients of >0.9 were aimed for (Table 1). However, a low RMSE for calibration and prediction, together with a low number of LVs, was accomplished for the selected model. The deviation of the data from the target line in the measured versus predicted plot (Figure 5A) was smaller for runs I and II than for cultivation run III. Nevertheless, it was possible to accept the model since 99.4% of the *x*-data variance and 99.3% of the *y*-data variance was captured. The scores plot showed that the EEMs of all four fermentations behaved similar after preprocessing of the raw data by background subtraction (Figure 5B). The three cultivations which were included in the model calibration covered final specific activities from 276 to 426 pmol·min^{-1}·µg^{-1}. Knowing that the cell factory showed a certain biological diversity supports the need for in situ monitoring, since the offline analytics are too time-intensive, and even small variations of the process due to the medium storage or the use of a different enzyme lots had an impact on the cell factory. The specific activity of the enzyme increased strongly within the two hours after booster and glucoamylase addition (Figure 5C–F), as it was also observed for the soluble HisDapGalNAcT2 to dry matter ratio. The results showed that the extraction of soluble protein from the cell lysate followed by the activity assay was prone to errors in the case of low cell concentrations and, therefore, also low enzyme concentrations. Outliers were predominantly suspected for samples with OD$_{600}$ values < 5. Consequentially only specific activities determined for samples with OD$_{600}$ values > 5 were used for model calibration. It was assumed, that at this point the chaperones were expressed in a sufficient quantity to obtain the active conformation of the glycosyltransferase. This resulted in a limitation of the selected PLS model, which was already suspected due to the low correlation coefficients. The accuracy of the calibration models is always dependent on the accuracy of assay or the method used for the analysis of the respective response variable. Therefore, the prediction of difficult-to-access process parameters remains a challenge. However, the data supported the earlier described assumption of a slow growth rate and preventing the cells from a metabolism overflow supported the constant accumulation of soluble glycosyltransferase in the cytoplasm [7]. Fluorescence EEMs recorded during cultivation run IV were used to test the soft sensor regarding the online predictability of the cell factory performance. A specific activity of 280 pmol·min^{-1}·µg^{-1} was determined for the HisDapGalNAcT2 at the end of cultivation run IV. Furthermore, a specific activity of 265.5 ± 3.2 pmol·min^{-1}·µg^{-1} was predicted based on the PLS model and the EEM recorded during the last 30 min of the cultivation (Figure 5F). This was a promising result, since a RSMEP of 33.5 pmol·min^{-1}·µg^{-1} and a RMSEC of only 30.7 pmol·min^{-1}·µg^{-1} was calculated for this soft sensor (Table 1).

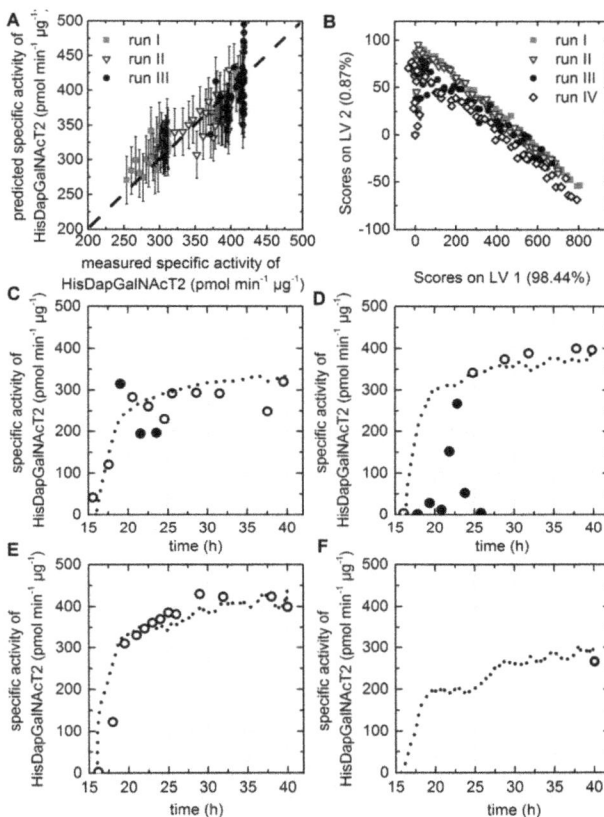

Figure 5. PLS model for the prediction of the specific activity of HisDapGalNAcT2. (**A**) The predicted versus reference plot of the PLS Model with three LVs. (**B**) The score values of all EEMs calculated for LV1 and LV2 are shown in the scores plot. (**C–F**) Offline (circles) and predicted (dashed line) values based on the PLS model are shown for cultivation I–IV. Only samples taken from the bioreactor with a determines $OD_{600} > 5.0$ were used as input data for the model, since a high number of possible outliers (closed circles) were identified for samples with lower OD_{600} values.

4. Conclusions

The specific activity measurements indicated that the complex and interlinked expression of glycosyltransferase and chaperones was extremely sensitive to process variations. The results suggested that the slower growth of the recombinant *E. coli* SHuffle® T7 strain in EnPresso® B medium slowed down the protein expression and presumably enabled the chaperone-mediated folding and disulfide-bound formation [13]. It was feasible to set up a reliable in situ monitoring for OD_{600} and acetate concentrations. To provide a PLS model for the prediction of glucose concentrations was challenging. The glucose release and therefore the glucose concentrations in the medium were not only dependent on the glucose consumption by the cells but also on the amylase activity. Furthermore, the development of a soft sensor for in situ prediction of the soluble protein content in the cells and the specific activity of HisDapGalNAcT2 was complex. One drawback in this context was that the specific activity of the enzyme had to be measured after purification from an *E. coli* cell lysate prior to PLS modeling. The results indicate that the use of more datasets might be beneficial for the calibration of such PLS models. Moreover, an improved assay for the determination of the specific activity of

11

Bioengineering **2016**, *3*, 32

HisDapGalNAcT2 might facilitate the model calibration. Nevertheless, the study pointed out that time-consuming and costly offline analysis might be rendered unnecessary for complex expression systems in the near future.

Supplementary Materials: The following are available online at http://www.mdpi.com/2306-5354/3/4/32/s1, Figure S1: *E. coli* Shuffle® T7 cell factory LB-medium fed-batch process. OD_{600}, as well as glucose and acetate, concentrations were measured offline. The glucose target concentration for the feed was $1\,g \cdot L^{-1}$. The cell factory was induced following the same protocol as described for the EnPresso® B medium. Only negligible amounts of glycosyltransferase were formed and the fermentation was stopped after increased inclusion body accumulation was observed and the *E. coli* cell morphology changed.

Acknowledgments: This research was supported by the German Federal Ministry of Education and Research (Grant No. 0315342A) and by the Cooperative Research Training Group Pharmaceutical Biotechnology stated by the Postgraduate Scholarships Act of the Ministry for Science, Research and Arts of the federal state government of Baden-Württemberg, Germany. In addition the authors are grateful to Lloyd W. Ruddock for providing the project with essential plasmids.

Author Contributions: Karen Schwab (K.S.) and Jennifer Lauber (J.L.) contributed to the conception and design of the study, performed experimental work and participated in data analysis. Friedemann Hesse (F.H.) contributed to the conception and design of the study. K.S., J.L. and F.H. made contributions to the writing of the manuscript. All authors read and approved the final manuscript.

Conflicts of Interest: The authors declare no conflict of interest.

References

1. Lobstein, J.; Emrich, C.A.; Jeans, C.; Faulkner, M.; Riggs, P.; Berkmen, M. Shuffle, a novel *Escherichia coli* protein expression strain capable of correctly folding disulfide bonded proteins in its cytoplasm. *Microb. Cell Fact.* **2012**, *11*, 753. [CrossRef] [PubMed]
2. Wong, J.W.; Ho, S.Y.; Hogg, P.J. Disulfide bond acquisition through eukaryotic protein evolution. *Mol. Biol. Evol.* **2011**, *28*, 327–334. [CrossRef] [PubMed]
3. Stock, J.; Rauch, B.; Roseman, S. Periplasmic space in salmonella typhimurium and *Escherichia coli*. *J. Biol. Chem.* **1977**, *252*, 7850–7861. [PubMed]
4. Ritz, D.; Lim, J.; Reynolds, C.M.; Poole, L.B.; Beckwith, J. Conversion of a peroxiredoxin into a disulfide reductase by a triplet repeat expansion. *Science* **2001**, *294*, 158–160. [CrossRef] [PubMed]
5. Stewart, E.J.; Åslund, F.; Beckwith, J. Disulfide bond formation in the *Escherichia coli* cytoplasm: An in vivo role reversal for the thioredoxins. *EMBO J.* **1998**, *17*, 5543–5550. [CrossRef] [PubMed]
6. Lauber, J.; Handrick, R.; Leptihn, S.; Dürre, P.; Gaisser, S. Expression of the functional recombinant human glycosyltransferase galnact2 in *Escherichia coli*. *Microb. Cell Fact.* **2015**, *14*, 3. [CrossRef] [PubMed]
7. Hortsch, R.; Weuster-Botz, D. Growth and recombinant protein expression with *Escherichia coli* in different batch cultivation media. *Appl. Microbiol. Biotechnol.* **2011**, *90*, 69–76. [CrossRef] [PubMed]
8. Hoffmann, F.; van den Heuvel, J.; Zidek, N.; Rinas, U. Minimizing inclusion body formation during recombinant protein production in *Escherichia coli* at bench and pilot plant scale. *Enzym. Microb. Technol.* **2004**, *34*, 235–241. [CrossRef]
9. Luchner, M.; Striedner, G.; Cserjan-Puschmann, M.; Strobl, F.; Bayer, K. Online prediction of product titer and solubility of recombinant proteins in *Escherichia coli* fed-batch cultivations. *J. Chem. Technol. Biotechnol.* **2015**, *90*, 283–290. [CrossRef]
10. Cruz, M.V.; Sarraguça, M.C.; Freitas, F.; Lopes, J.A.; Reis, M.A.M. Online monitoring of P(3HB) produced from used cooking oil with near-infrared spectroscopy. *J. Biotechnol.* **2015**, *194*, 1–9. [CrossRef] [PubMed]
11. Lindemann, C.; Marose, S.; Nielsen, H.O.; Scheper, T. 2-dimensional fluorescence spectroscopy for on-line bioprocess monitoring. *Sens. Actuators B Chem.* **1998**, *51*, 273–277. [CrossRef]
12. Mazarevica, G.; Diewok, J.; Baena, J.R.; Rosenberg, E.; Lendl, B. On-line fermentation monitoring by mid-infrared spectroscopy. *Appl. Spectrosc.* **2004**, *58*, 804–810. [CrossRef] [PubMed]
13. Jain, G.; Jayaraman, G.; Kökpinar, Ö.; Rinas, U.; Hitzmann, B. On-line monitoring of recombinant bacterial cultures using multi-wavelength fluorescence spectroscopy. *Biochem. Eng. J.* **2011**, *58–59*, 133–139. [CrossRef]
14. Schwab, K.; Amann, T.; Schmid, J.; Handrick, R.; Hesse, F. Exploring the capabilities of fluorometric online monitoring on cho cell cultivations producing a monoclonal antibody. *Biotechnol. Prog.* **2016**. [CrossRef] [PubMed]

15. Teixeira, A.P.; Portugal, C.A.M.; Carinhas, N.; Dias, J.M.L.; Crespo, J.P.; Alves, P.M.; Carrondo, M.J.T.; Oliveira, R. In situ 2D fluorometry and chemometric monitoring of mammalian cell cultures. *Biotechnol. Bioeng.* **2009**, *102*, 1098–1106. [CrossRef] [PubMed]
16. Mercier, S.M.; Diepenbroek, B.; Dalm, M.C.; Wijffels, R.H.; Streefland, M. Multivariate data analysis as a pat tool for early bioprocess development data. *J. Biotechnol.* **2013**, *167*, 262–270. [CrossRef] [PubMed]
17. Wold, S.; Esbensen, K.; Geladi, P. Principal component analysis. *Chemom. Intell. Lab. Syst.* **1987**, *2*, 37–52. [CrossRef]
18. Wold, S.; Sjöström, M.; Eriksson, L. PLS-regression: A basic tool of chemometrics. *Chemom. Intell. Lab. Syst.* **2001**, *58*, 109–130. [CrossRef]
19. Duggan, D.E.; Bowman, R.L.; Brodie, B.B.; Udenfriend, S. A spectrophotofluorometric study of compounds of biological interest. *Arch. Biochem. Biophys.* **1957**, *68*, 1–14. [CrossRef]
20. Marose, S.; Lindemann, C.; Scheper, T. Two-dimensional fluorescence spectroscopy: A new tool for on-line bioprocess monitoring. *Biotechnol. Prog.* **1998**, *14*, 63–74. [CrossRef] [PubMed]
21. Haack, M.B.; Eliasson, A.; Olsson, L. On-line cell mass monitoring of saccharomyces cerevisiae cultivations by multi-wavelength fluorescence. *J. Biotechnol.* **2004**, *114*, 199–208. [CrossRef] [PubMed]
22. Hisiger, S.; Jolicoeur, M. A multiwavelength fluorescence probe: Is one probe capable for on-line monitoring of recombinant protein production and biomass activity? *J. Biotechnol.* **2005**, *117*, 325–336. [CrossRef] [PubMed]
23. Ganzlin, M.; Marose, S.; Lu, X.; Hitzmann, B.; Scheper, T.; Rinas, U. In situ multi-wavelength fluorescence spectroscopy as effective tool to simultaneously monitor spore germination, metabolic activity and quantitative protein production in recombinant aspergillus niger fed-batch cultures. *J. Biotechnol.* **2007**, *132*, 461–468. [CrossRef] [PubMed]
24. Rossi, D.M.; Solle, D.; Hitzmann, B.; Ayub, M.A.Z. Chemometric modeling and two-dimensional fluorescence analysis of bioprocess with a new strain of klebsiella pneumoniae to convert residual glycerol into 1,3-propanediol. *J. Ind. Microbiol. Biotechnol.* **2012**, *39*, 701–708. [CrossRef] [PubMed]
25. Hantelmann, K.; Kollecker, M.; Hull, D.; Hitzmann, B.; Scheper, T. Two-dimensional fluorescence spectroscopy: A novel approach for controlling fed-batch cultivations. *J. Biotechnol.* **2006**, *121*, 410–417. [CrossRef] [PubMed]
26. Ödman, P.; Johansen, C.L.; Olsson, L.; Gernaey, K.V.; Lantz, A.E. On-line estimation of biomass, glucose and ethanol in saccharomyces cerevisiae cultivations using in-situ multi-wavelength fluorescence and software sensors. *J. Biotechnol.* **2009**, *144*, 102–112. [CrossRef] [PubMed]
27. Van Dat Nguyen, F.H.; Salo, K.E.; Enlund, E.; Zhang, C.; Ruddock, L.W. Pre-expression of a sulfhydryl oxidase significantly increases the yields of eukaryotic disulfide bond containing proteins expressed in the cytoplasm of E. coli. *Microb. Cell Fact.* **2011**, *10*, 1. [CrossRef] [PubMed]
28. Wise, B.M.; Gallagher, N.B.; Bro, R.; Shaver, J.M. *PLS Toolbox 3.0*; Eigenvector Research Inc.: Manson, WA, USA, 2003; Volume 171.
29. Wold, S. Chemometrics; what do we mean with it, and what do we want from it? *Chemom. Intell. Lab. Syst.* **1995**, *30*, 109–115. [CrossRef]
30. Panula-Perälä, J.; Siurkus, J.; Vasala, A.; Wilmanowski, R.; Casteleijn, M.G.; Neubauer, P. Enzyme controlled glucose auto-delivery for high cell density cultivations in microplates and shake flasks. *Microb. Cell Fact.* **2008**, *7*, 31. [CrossRef] [PubMed]
31. Diaz Ricci, J.C.; Hernández, M.E. Plasmid effects on Escherichia coli metabolism. *Crit. Rev. Biotechnol.* **2000**, *20*, 79–108. [CrossRef] [PubMed]

bioengineering

MDPI

Article

Estimating Extrinsic Dyes for Fluorometric Online Monitoring of Antibody Aggregation in CHO Fed-Batch Cultivations

Karen Schwab * and Friedemann Hesse

Institute of Applied Biotechnology, Biberach University of Applied Science, 77781 Biberach, Germany; hesse@hochschule-bc.de
* Correspondence: karen.schwab@gmail.com; Tel.: +49-7351-582-441

Academic Editor: Christoph Herwig
Received: 24 June 2017; Accepted: 17 July 2017; Published: 24 July 2017

Abstract: Multi-wavelength fluorescence spectroscopy was evaluated in this work as tool for real-time monitoring of antibody aggregation in CHO fed-batch cultivations via partial least square (PLS) modeling. Therefore, we used the extrinsic fluorescence dyes 1-anilinonaphthalene-8-sulfonate (ANS), 4,4′-bis-1-anilinonaphthalene-8-sulfonate (Bis-ANS), or Thioflavin T (ThT) as medium additives. This is a new application area, since these dyes are commonly used for aggregate detection during formulation development. We determined the half maximum inhibitory concentrations of ANS (203 ± 11 µmol·L^{-1}), Bis-ANS (5 ± 0.5 µmol·L^{-1}), and ThT (3 ± 0.2 µmol·L^{-1}), and selected suitable concentrations for this application. The results showed that the emission signals of non-covalent dye antibody aggregate interaction superimposed the fluorescence signals originating from feed medium and cell culture. The fluorescence datasets were subsequently used to build PLS models, and the dye-related elevated fluorescence signals dominated the model calibration. The soft sensors based on ANS and Bis-ANS signals showed high predictability with a low error of prediction (1.7 and 2.3 mg·mL^{-1} aggregates). In general, the combination of extrinsic dye and used concentration influenced the predictability. Furthermore, the ThT soft sensor indicated that the intrinsic fluorescence of the culture might be sufficient to predict antibody aggregation online.

Keywords: antibody aggregation; ANS; bioprocess monitoring; Bis-ANS; fluorescence spectroscopy; CHO

1. Introduction

One objective of process analytical technology (PAT) is to improve process understanding at any process step by identifying the related critical process parameters (Food and Drug Administration 2004). One possibility to gain a better understanding of cultivation processes in bioreactors is the implementation of online monitoring based on spectroscopic soft sensors [1]. However, spectroscopic data cannot be interpreted by univariate statistical techniques, which are mostly applied to access industrial data sets [2]. Multivariate data analysis has to be used to extract the required information from the huge dataset [3,4]. The most common approach is to use partial least square regression (PLSR) to interpret the data. The underlying algorithm is capable of extracting a linear dependency of an x-dataset and the corresponding y-dataset [5]. Offline measured response variables such as substrate, metabolites, and viable cell concentration can be used as y-dataset. The x-dataset mostly consists of large, noisy, and highly redundant data recorded by the spectroscopic sensors. The resulting PLSR model can be used for online and real-time prediction of the respective response variables. Among the available spectroscopic techniques [6–8], 2D fluorescence spectroscopy (2DFS) proved to be a reliable method which allowed online monitoring of viable cell concentrations as well as product and metabolite concentrations in mammalian cell batch and fed-batch cultivations with non-fluorescent feed [9]. Ohadi et al. (2014) described that it is possible to predict key process parameters at-line in Chinese hamster ovary cell (CHO) shake flask cultures [10]. Schwab et al. (2016) proved that it was possible to predict fluorescent and non-fluorescent key process parameters online in CHO fed-batch cultivations with highly fluorescent feed medium. So far, however, none of the existing studies have focused on online monitoring of product quality parameters during upstream processing (USP). One example of an important quality parameter which is reported to be an issue throughout the whole monoclonal antibody (mAb) production process is product aggregation [11]. MAb aggregates are known to reduce drug performance and can cause anaphylactic reactions [12]. It was shown by various groups that mAb aggregation can already occur during USP, since aggregation can be influenced by medium components, ionic strength, pH control, and temperature shift [13–15]. Paul et al. (2015) analyzed mAb aggregates directly in CHO cell culture supernatant via size exclusion chromatography, and reported roughly 77% aggregated mAb [16]. A high aggregate content during USP might eventually reduce the overall yield and increase the burden on downstream processing. The intrinsic fluorescence of proteins is based on the aromatic amino acids tryptophan, tyrosine, and phenylalanine, and Ohadi et al. (2015) proved that it is possible to use the intrinsic fluorescence of the mAb in combination with chemometric modeling in order to identify aggregate levels in purified protein samples at-situ [17,18]. Therefore, fluorometric online monitoring of mAb aggregation in CHO fed-batch cultivations should be possible in principle. However, it was suspected that the fluorescence signals of medium, feed medium, and cells could possibly outbalance the emission signal changes related to mAb aggregate formation. Therefore, the fluorescence dyes 4,4′-bis-1-anilinonaphthalene-8-sulfonate (Bis-ANS) and its monomeric analogue 1-anilinonaphthalene-8-sulfonate (ANS) along with thioflavin T (ThT) were used in this study. These extrinsic dyes show low fluorescence signals in aqueous solution, but the fluorescence signals increase upon non-covalent binding to partially unfolded or aggregated proteins [16,19]. Extrinsic dyes are usually applied as fluorescent markers in tissue sections, protein purification, and formulation development [20–23]. However, Paul et al. (2015) have already applied ANS and Bis-ANS for the detection of aggregated mAb in cell-free CHO culture supernatant. They reported that specific regions of the recorded excitation emission matrices (EEM) showed increased fluorescence signals which were directly related to increased aggregate concentrations [16]. Furthermore, these dyes can be solved in water instead of ethanol or DMSO like Congo red, Nile red, and DCVJ [23]. Nevertheless, none of these dyes have been used thus far as additive to mammalian cell cultivations in order to measure protein aggregation.

In this context, the current study aims to demonstrate that 2DFS in combination with extrinsic fluorescence dyes can be used for real-time monitoring of mAb aggregation during CHO fed-batch cultivations. Therefore, half maximum inhibitory concentrations (IC50) had to be determined for the extrinsic fluorescence dyes ANS, Bis-ANS, and ThT in CHO cultivations. The dyes created additional emission signals in the EEM which were used to build PLSR models for the prediction of mAb aggregate concentrations. The challenge was to estimate the concentrations for each dye that allowed the calibration of a reliable PLSR model. The resulting soft sensors were compared regarding their predictive capability and susceptibility to disturbance by feed additions. Based on the results, we evaluated which extrinsic dye was most suitable and can be used for real-time monitoring of mAb aggregation in CHO fed-batch cultivations.

2. Materials and Methods

2.1. Cell Line, Medium, and Culture Conditions

The CHO DG44 cell line expressing an aggregation-prone mAb was used as model system. The cells were seeded at $2–4 \times 10^5$ mL^{-1} in SFM4CHO medium (GE Healthcare, Chicago, IL, USA), incubated at 37 °C and 5% CO_2, and passaged every 3–4 days. Glucose and L-glutamine (Life

Technologies, Carlsbad, CA, USA) were added to the medium upon usage. All chemicals were ordered from Roth (Karlsruhe, Germany) if not stated otherwise.

2.2. IC50 Experiments

Stock solutions of 3 mmol·L^{-1} ANS, 1 mmol·L^{-1} Bis-ANS, and 2 mmol·L^{-1} ThT (Sigma-Aldrich, Taufkirchen, Germany) in SFM4CHO were prepared and sterile filtered with 0.2 µm (Phenomenex, Torrance, CA, USA). The following dye concentration ranges were used for toxicity evaluation: (a) 0–400 µmol·L^{-1} ANS; (b) 0–80 µmol·L^{-1} Bis-ANS; and (c) 0–20 µmol·L^{-1} ThT. Exponentially growing CHO cells were spun down, diluted in fresh medium, and seeded with a final cell concentration of 4×10^5 mL^{-1} ($n = 3$) in 24-well plates with a final volume of 2 mL (Greiner Bio-one, Frickenhausen, Germany). The fluorescence dyes were directly supplied to the wells together with the medium containing 5 g·L^{-1} glucose and 4 mmol·L^{-1} L-glutamine. The plates were covered with breath seals (Greiner Bio-one, Frickenhausen, Germany) in addition to the lit and incubated in a shaker with a maximum deflection of 4 mm (Kuhner, Basel, Switzerland). The final viable cell concentrations and viabilities were determined after 72 h cultivation using a FACSCalibur flow cytometer (DB Bioscience, San Jose, CA, USA), following the propidium iodide staining protocol proposed by Cummings and Schnellmann [24].

2.3. Fed-Batch Cultures with Fluorescence Dyes

Three fed-batch cultivations were performed for each dye, and the details are listed in Table 1. The fed-batch cultivations were inoculated with a cell concentration of 10×10^5 mL^{-1}. A 2-L benchtop bioreactor BIOSTAT$^{®}$ Bplus (Sartorius, Göttingen, Germany) was used, and the temperature was kept constant at 37 °C. The dissolved oxygen saturation was kept at 60% through sparging with an aeration rate of 0.25 vvm, while pH 7 was maintained through CO_2 or 1 M NaOH addition. The culture was stirred at 100 rpm, and the glucose target concentration for feeding was set between 0.8–1.2 g·L^{-1}. Cell Boost 6 (4% w/v) (GE Healthcare, Chicago, IL, USA) was solved in deionized water and supplemented to the bioreactor continuously as glucose feed. The feed rate was adjusted depending on the glucose consumption rates. Likewise, L-glutamine was diluted in SFM4CHO (100 mmol·L^{-1}) and added as a bolus feed when required to keep the concentrations $\geq$0.8 mmol·L^{-1}. The extrinsic dyes were added to the bioreactor 67 ± 4 h after the inoculation, with a final concentration of either 100 µmol·L^{-1} ANS, 2 µmol·L^{-1} Bis-ANS, or 2 µmol·L^{-1} ThT. Glucose and L-glutamine feeds contained the same fluorescence dye concentration as the cell culture.

Table 1. Culture conditions for all fed-batch fermentations expressing a full-size monoclonal antibody in the presence of extrinsic dyes. ANS: 1-anilinonaphthalene-8-sulfonate; Bis-ANS: 4,4′-bis-1-anilinonaphthalene-8-sulfonate; mAb: monoclonal antibody; Th T: thioflavin T.

Cultivation	Start Concentration		Cultivation Time (h)	Time of Dye Addition (h)	X_V max ($\times 10^6$ mL^{-1})	mAb (mg·L^{-1})	Aggregated mAb (mg·L^{-1})
	Glucose (g·L^{-1})	Glutamine (mmol·L^{-1})					
ANS I	1.63	1.26	208	64	5.06	95	60
ANS II	1.90	1.75	208	63	4.33	89	61
ANS III	2.84	2.03	212	74	3.38	88	59
Bis-ANS I	1.92	2.14	161	66	4.86	61	41
Bis-ANS II	2.37	1.93	165	69	4.55	57	40
Bis-ANS III	2.40	2.02	215	70	3.00	71	45
Th T I	1.99	2.08	204	63	4.86	58	42
Th T II	1.93	1.91	209	68	4.06	65	48
Th T III	2.84	2.03	214	67	3.67	59	43

2.4. Offline Analytics

The antibody concentration was determined with protein A HPLC using a POROS$^{®}$ 20 µm column (Thermo Fisher Scientific, Waltham, MA, USA) and the UltiMate 3000 system (Thermo Fisher

Scientific, Waltham, MA, USA). The mAb aggregate concentration in the cell culture samples was determined with SE-HPLC using the Agilent 1100 system (Agilent Technologies, Santa Clara, CA, USA). A MAbPac SEC-1 column (Thermo Fisher Scientific, Waltham, MA, USA) was used, following the method described by Paul et al. (2014) for the direct determination of mAb aggregate concentration in cell culture samples [25]. L-glutamine and glucose concentrations were determined enzymatically with the KonelabTM 20 XT (Thermo Fisher Scientific, Waltham, MA, USA) using the L-glutamine kit (Thermo Fisher Scientific, Waltham, MA, USA) and the Glucose HK kit (Thermo Fisher Scientific, Waltham, MA, USA), respectively. Cells were counted using a Cedex XS analyzer (Roche, Basel, Switzerland) and trypan blue exclusion.

2.5. Online Data Collection

The multi-wavelength EEMs were recorded with the BioView® sensor (Delta, Hørsholm, Denmark), using a fiber optic assembly for the 19 mm port. A gain of 1300 and a measurement interval of 15 min were set, and the glass vessels were covered with blackout material. Excitation wavelengths from 270–550 nm and emission wavelengths ranging from 310–590 nm in 20 nm steps were used to record the EEMs. The resulting fluorescence datasets were exported vectorized into two-way arrays containing 120 excitation/emission wavelength pairs ($\lambda_{ex/em}$) per EEM. A detailed description of the instrument can be found elsewhere [26–28].

2.6. Chemometric Modeling and Data Preprocessing

All chemometric methods were performed using MatLab version 8.4.0 (MathWorks, Natick, MA, USA) in combination with the PLS-toolbox version 7.9.5 (Eigenvector Research Ing., Manson, WA, USA) and a detailed description of the chemometric method can be found elsewhere [5,29]. The EEMs of all fed-batch cultivations were preprocessed with multiplicative signal correction (MSC). This was followed by an external parameter orthogonalization (EPO) filtering method with one principal component [30] (Table 2) for cultivations containing ANS and Bis-ANS. No further preprocessing was applied to datasets from cultivations containing ThT. The SIMPLS algorithm was used for PLSR modeling. EEMs and corresponding offline measured aggregate concentrations were used as input for the model generation. Only EEMs and corresponding offline data recorded after dye addition were used for PLSR modeling. The samples were randomly split into a calibration and a validation dataset (66–34%) using the onion method. The resulting models were selected with the aim of minimizing the RSME of calibration (RMSEC), cross-validation (RMSECV), and prediction (RMSEP) with the preferably lowest possible number of latent variables (LV) [29]. MAb aggregate concentrations were predicted based on all EEMs recorded during the cultivation and the selected chemometric models. Quality parameters of the selected models are listed in Table 2.

Table 2. Partial least square regression (PLSR) modeling results for the prediction of aggregated mAb concentrations. EPO: External parameter orthogonalization; MSC: multiplicative signal correction.

			Calibration		Cross-Validation		Prediction	
Extrinsic Dye	**Preprocessing**	**LV**	R^2_{cal}	**RMSEC** [mg·mL^{-1}]	R^2_{CV}	**RMSECV** [mg·mL^{-1}]	R^2_P	**RMSEP** [mg·mL^{-1}]
ANS	MSC (mean) EPO (1 PC)	5	0.97	1.1	0.94	2.6	0.97	1.7
Bis-ANS	MSC (mean) EPO (1 PC)	5	0.96	1.9	0.89	3.2	0.92	2.3
ThT	MSC (mean)	6	0.92	2.2	0.85	3.0	0.85	3.1

3. Results

3.1. IC50 Experiments

CHO cells were seeded in microtiter plates and cultivated in the presence of ANS, Bis-ANS, and ThT. Viable cell concentrations and viabilities were determined after a cultivation time of 72 h using flow cytometry and propidium iodide staining. The viable cell concentrations were plotted against the logarithmic concentration of the particular dye. Sigmoid curves were fitted to the data points, and the respective IC50s were calculated (Figure 1A–C). ANS with a calculated IC50 of 203 ± 11 μmol·L^{-1} was less toxic than the other dyes. In comparison, an IC50 of 5 ± 0.5 μmol·L^{-1} was calculated for its dimeric analogue Bis-ANS, and an even lower IC50 of 3 ± 0.2 μmol·L^{-1} was calculated for ThT. The sigmoid curve fit was also used for the viability data (Figure 1D–F). This made the comparison of cell concentrations and cell viabilities at half maximum inhibitory dye concentration possible. A viability of 80% was calculated for the IC50 concentration of ANS. Furthermore, a viability of 91% was determined for an IC50 concentration of Bis-ANS or ThT. This showed that the cell growth was already reduced compared to the control experiment without dye, but the viability was not affected in the same way.

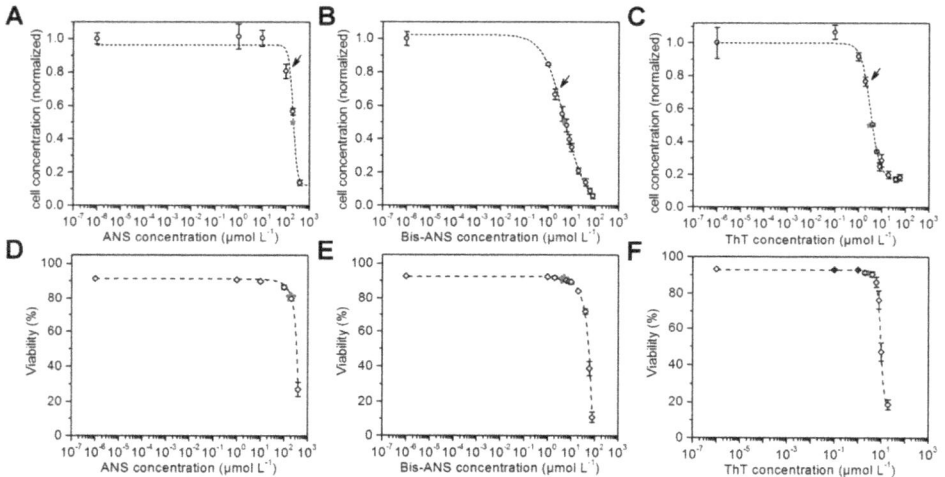

Figure 1. Cell concentration after 72 h cultivation in the presence of ANS (IC50 of 203 ± 11 μmol·L^{-1}), Bis-ANS (IC50 of 5 ± 0.5 μmol·L^{-1}) and ThT (IC50 of 3 ± 0.2 μmol·L^{-1}) (**A–C**) and the corresponding viabilities (**D–F**) are shown. A sigmoid curve fit was performed in order to determine the corresponding IC50s (indicated by red stars). Arrows indicate the dye concentrations that were used later in the fed-batch processes.

3.2. Online Monitoring Based on Extrinsic Fluorescence

CHO cells expressing an aggregation-prone mAb were cultivated in 2-L bioreactors in fed-batch mode containing 100 μmol·L^{-1} ANS, 2 μmol·L^{-1} Bis-ANS, or 2 μmol·L^{-1} ThT, respectively, as medium additives. The cultivations were inoculated with high cell concentrations, and the extrinsic dyes were added after 67 ± 4 h in order to ensure antibody concentrations $\geq$20 mg·L^{-1}. The selected dye concentrations were expected to be high enough to create additional recognizable fluorescence signals in the EEM. It was furthermore anticipated that these additional emission signals facilitate the calibration of PLSR models for the online prediction of mAb aggregate concentrations in CHO fed-batch cultivations. Culture conditions as well as dye and substrate concentrations in medium and feed are stated in Table 1. Samples were drawn from the bioreactor, and product concentrations and aggregate levels were determined following the method described by Paul et al. (2014) [25]. The PLSR models were calibrated with data recorded after dye addition. The final models were selected based on the quality parameters listed in Table 2 and their ability to predict the response variable based on the additionally recorded EEMs during cultivation. The trajectories on the scores plot of cultivations that contained either ANS or Bis-ANS showed a straight evolvement over the first two latent variables

(LVs) (Figure 2A,B). Contrary to this, trajectories of the ThT cultivations indicated that besides the emission signals related to ThT, signals derived from CHO cells, medium, and feed also had an impact on the model (Figure 2C). This was demonstrated through the differences between the trajectories of the three fed-batch cultivations. It was not possible to increase the model quality through extensive preprocessing. Furthermore, outliers were identified in the scores plot of cultivation run III containing ThT. They were related to interfering light at the end of the cultivation (Figure 2C).

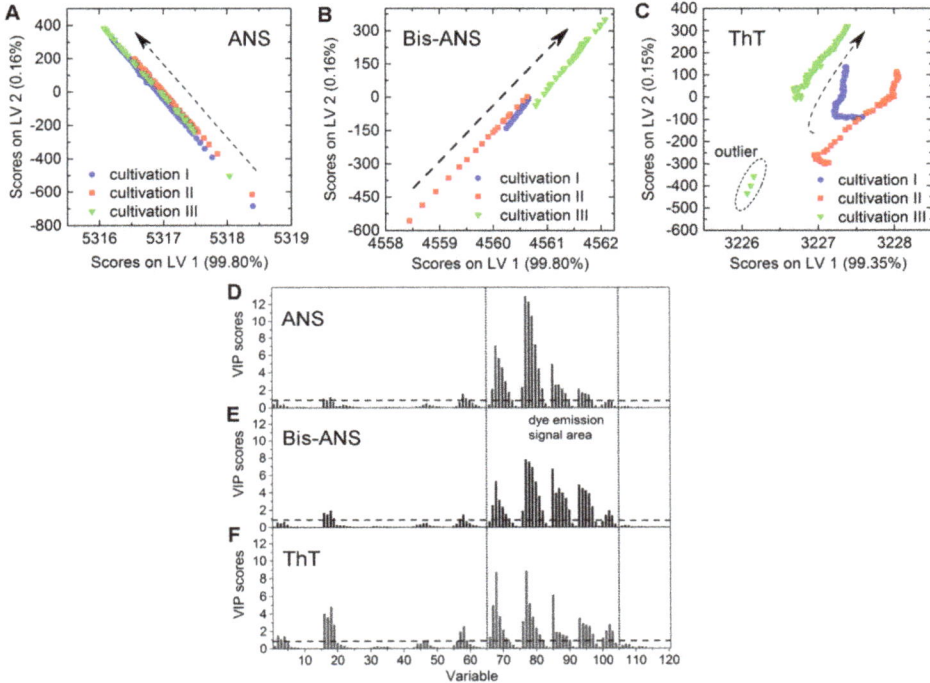

Figure 2. Scores plots of the PLS models for the prediction of aggregated mAb containing either (**A**) 100 μmol·L^{-1} ANS; (**B**) 2 μmol·L^{-1} Bis-ANS; or (**C**) 2 μmol·L^{-1} ThT. Dashed arrows indicate the development of the trajectories over cultivation time (only every 20th data point is shown). Furthermore, corresponding variable importance in projection (VIP) scores plots of the respective PLSR models are given in (**D–F**), respectively. LV: latent variable.

A direct comparison of the different PLS models can be done by comparing the respective variable importance in projection (VIP) scores plots. The importance of each wavelength pair for the projection in the model is shown by the VIP scores plots. Each recorded EEM can be vectorized into 120 wavelength pairs. Accordingly, the resulting PLSR model consisted of 120 variables. Variables with VIP scores close to 1 or higher are considered as important for the model [27,31]. The highest VIP scores in all three PLSR models were determined for the area between variable 66–105 (λ_{ex} 370–450 nm and λ_{em} 410–590 nm) of the VIP scores plot (Figure 2D–F). Furthermore, a second area between variable 16 and 28 (λ_{ex} 290 nm and λ_{em} 350–570 nm) showed VIP values close to one. More than 99% of the x- and y-variance were captured by 1 LVs of the PLSR models computed for cultivations containing ANS or Bis-ANS. However, R^2_{cal} and R^2_{CV} improved with the implementation of more LVs in both cases. Therefore, five LVs were selected for both models in order to meet the above proposed criteria of low RMSE's, R^2's close to 1, together with the lowest possible number of LVs. Over 99% of the x-variance was already explained by LV1 of the ThT model, but only 95.2% of the variance in the y-dataset was captured. Furthermore, the R^2_{cal} of this model improved with five additional LVs from <0.7 to 0.92. In general, the predicted and offline values of all cultivations and all models showed a close fit to the target line of the predicted versus measured plots (Figure 3). The PLSR model calculated for cultivations containing ANS showed the best correlations between offline values and predicted mAb aggregate concentrations (Figure 3A). An RMSEP of 1.7 mg·L^{-1} aggregated mAb was calculated for this model (Table 2). The Bis-ANS model performed in a similar way, but a higher RMSEP of 2.3 mg·L^{-1} mAb aggregates was calculated. Furthermore, an RMSEP of 3.1 mg·L^{-1} mAb aggregates was calculated for the ThT model. EEMs not included in the model calibration were used in a next step as input data for the aggregate prediction. Overall, a good correlation between measured and predicted values was observed (Figure 3), with some exception for Bis-ANS cultivation III and ThT cultivation III. For ThT cultivation III, the prediction assumed a rapid increase in product aggregation prior to harvest. For Bis-ANS cultivation III, the prediction indicated an increase in mAb aggregates, whereas the offline measured values suggested decreasing values.

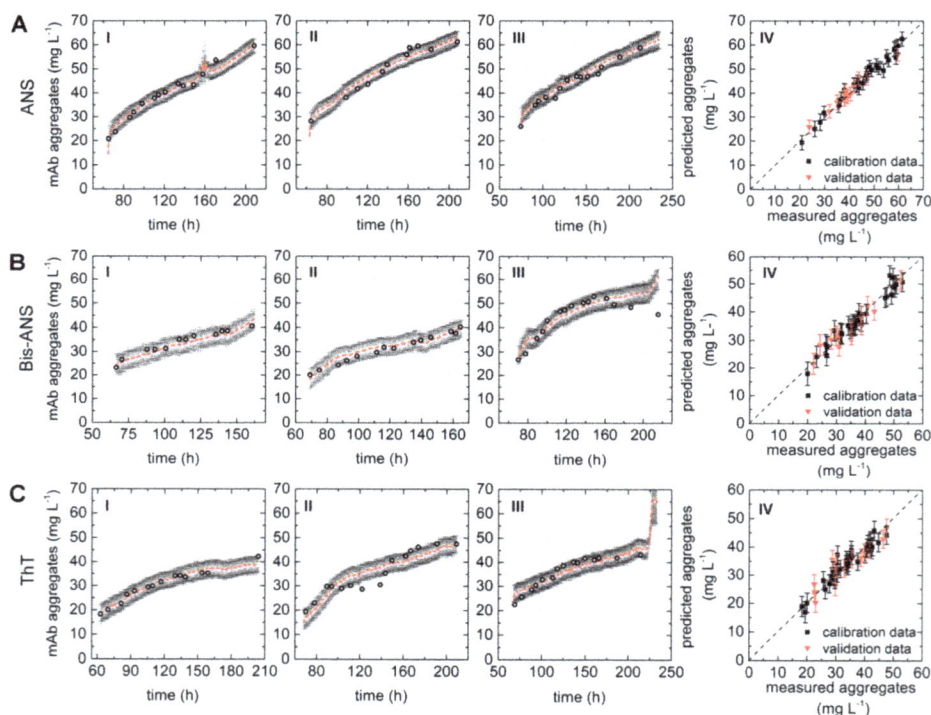

Figure 3. Correlation plots as a practical test by comparing offline values and values predicted based on the PLSR models calculated with the data recorded during three fed-batch cultivations (**I–III**) containing (**A**) 100 µmol·L^{-1} ANS; (**B**) 2 µmol·L^{-1} Bis-ANS; and (**C**) 2 µmol·L^{-1} ThT. Offline measured measurements (circles), predicted values (red), and corresponding confidence limits of the prediction based on the calculated errors (black). Plot IV (**A–C**) shows the predicted versus reference plots of the resulting PLSR models.

4. Discussion

4.1. Fluorescence Dye Toxicity

Fluorescence dyes have so far been mainly used in protein purification and formulation development [32–35]. Their influence on cell growth and viability had to be investigated first in order to determine suitable concentrations of ANS, Bis-ANS, and ThT that can be added to CHO fed-batch cultures. Suitable dye concentrations were selected for further experiments based on the determined IC50s as well as on concentrations reported in literature. For example, Hawe et al. (2010) [35] studied the ability of 5 μmol·L^{-1} Bis-ANS to detect aggregation in commercially available therapeutic antibodies formulations. A suitable concentration of 50 μmol·L^{-1} ANS for the determination of antibody monomers and oligomers was proposed by Franey et al. (2010) [34]. Amani et al. (2014) used 10 μmol·L^{-1} ThT to investigate the influence of detergents addition on antibody conformation in formulations. Finally, Paul et al. (2015) used 10 μmol·L^{-1} Bis-ANS and 100 μmol·L^{-1} ThT in their at situ study. Their proposed high-throughput compatible method was able to distinguish between different amounts of mAb aggregates in cell culture supernatant. Based on literature and the toxicity experiments, it was expected that concentrations of 100 μmol·L^{-1} ANS, 2 μmol·L^{-1} Bis-ANS, and 2 μmol·L^{-1} ThT would be high enough to generate a distinct additional signal in the fluorescence scans. It became evident during the study that the selection of suitable dye concentrations is important, especially for high titer processes. The potential aggregate concentration and the used dye and its concentration have to be aligned, and might have to be determined for different production processes accordingly.

4.2. Extrinsic Dye-Related Emission Signals

The areas in the EEMs (λ_{ex} 370–450 nm and λ_{em} 410–590 nm) with increased fluorescence due to non-covalent dye aggregate interaction were identified based on the VIP scores plots (Figure 2D–F). The respective fluorescence areas were in accordance with Hawe et al. (2008b) [23], who proposed the same wavelengths for the applied dyes. The experiments also showed that fluorescence dye addition not only added a new emission signal to the EEMs, but also intensified the fluorescence signals of the whole scan. This became evident through the direct comparison of the EEM before and after the addition of the extrinsic dye to the cell culture. In general, the overall fluorescence signal intensity increased, and the area with the highest emission signals was dependent on the dye and its concentration (ANS > Bis-ANS > ThT). The resulting PLSR models were built based on so-called primary effects, since the non-covalent interaction of mAb aggregates and respective dye creates a fluorescence signal which is directly related to the y-dataset of the model. Interactions of the fluorescence dye and protein contained in the medium cannot be completely excluded. However, we already proved in previous work that PLSR models can be built for the prediction of mAb aggregates in cell culture supernatant using the extrinsic dyes Bis-ANS and ThT [16]. The selected PLSR models were able to predict the aggregate concentrations based on the recorded EEMs. Furthermore, predicted and measured concentrations were closely aligned in this work (Figure 3) and in the previously published work of Paul et al. (2015) [16]. Therefore, it can be concluded that an interaction of dye and media proteins was either compensated in the PLSR model or it was rather negligible.

4.3. Chemometric Modeling

Fluorescence signals derived from non-covalent interactions between hydrophobic regions of the mAb aggregates and either ANS or Bis-ANS created high emission signal intensities over a broad range in the EEMs. These signals were successfully correlated via PLSR to the aggregated mAb concentration. The fluorescent signals created by these dyes superimposed the fluorescence of feed medium and cell culture in specific areas of the EEM, as discussed earlier. This implied that the fluorescent feed medium had no impact on the fluorescence signals relevant for mAb aggregates. Furthermore, the score values suggested that the fluorescence increase of certain wavelength pairs was directly related to the increasing aggregate concentration. As a result, the calibration of PLSR models for the prediction of aggregated mAb based on extrinsic signals was feasible. In contrast, ThT was identified to be more toxic than ANS or Bis-ANS. Therefore ThT was only used in a rather low concentration compared to Paul et al. (2015) [16], which resulted in low fluorescence signal intensities in the area of λ_{ex} 370–450 nm and λ_{em} 410–590 nm. In contrast, the same Bis-ANS concentration generated higher fluorescence intensities in the same EEM area. However, the dye-related fluorescence signals did not completely outshine the signals generated through the fluorescent feed medium and cell culture. Therefore, a second region in the VIP scores plot showed its relevance for the model generation. The area of λ_{ex} 290 nm and λ_{em} 350–570 nm showed VIP values close to one, which can be related to the intrinsic fluorescence of the total protein concentration (Figure 2). This supported the assumption that the chemometric model was based on a combination of extrinsic signals which can also be called primary effects related to the fluorescence signal of ThT and intrinsic signals, so-called secondary effects of the culture and medium. We furthermore proved in earlier work that the addition of fluorescent feed did not hinder model generation [27]. The scores plots and the predictions identified some EEM of Bis-ANS cultivation run III and ThT cultivation run III as possible outliers (Figure 3). One reason could be a biased prediction due to interfering light because of scattering effects [27,36]. Another explanation could be mAb aggregate precipitation during sample preparation. This was already seen by Paul et al. (2014) [25], who described that large aggregates can be formed. These higher molecular weight species might be removed from the sample through centrifugation or filtration. From this study, it can be concluded that ANS was evaluated as the most promising candidate for this application, since the PLSR model showed high predictive power. Furthermore, the result indicated that the dye concentration of 100 mmol·L^{-1} could even be reduced. However, it is not applicable to use extrinsic dyes as media components in commercial biopharmaceutical production processes, since this would trigger questions concerning the product safety. Nevertheless, in some cases it could be beneficial to use fluorescence dyes to generate dominant signals for the chemometric modeling of mAb aggregate concentrations. It could be used in scaled-down models during media and process development or during clone selection screenings. It could be furthermore beneficial for early process development stages or during troubleshooting in order to identify raw materials that are suspected to trigger protein aggregation. The use of extrinsic dyes for late-stage process development is not feasible. Therefore, whether the intrinsic fluorescence of cell culture, medium, and product can be used to set up soft sensors for the prediction of antibody aggregation must be evaluated in further work.

Acknowledgments: This research was supported by the German Federal Ministry of Education and Research (Grant No. 0315342A). In addition the authors are grateful to Rentschler Biotechnologie GmbH (Laupheim, Germany) for providing the mAb-producing CHO cell line.

Author Contributions: K.S. contributed to the design and the conception of the study, performed the experimental work and the data analysis. F.H. contributed to the design and the conception of the study. K.S. and F.H. wrote the manuscript.

Conflicts of Interest: The authors declare no commercial or financial conflict of interest.

References

1. Luttmann, R.; Bracewell, D.G.; Cornelissen, G.; Gernaey, K.V.; Glassey, J.; Hass, V.C.; Kaiser, C.; Preusse, C.; Striedner, G.; Mandenius, C.F. Soft sensors in bioprocessing: A status report and recommendations. *Biotechnol. J.* **2012**, *7*, 1040–1048. [CrossRef] [PubMed]
2. Mercier, S.M.; Diepenbroek, B.; Dalm, M.C.; Wijffels, R.H.; Streefland, M. Multivariate data analysis as a PAT tool for early bioprocess development data. *J. Biotechnol.* **2013**, *167*, 262–270. [CrossRef] [PubMed]
3. Glassey, J.; Gernaey, K.; Clemens, C.; Schulz, T.W.; Oliveira, R.; Striedner, G.; Mandenius, C.F. Process analytical technology (PAT) for biopharmaceuticals. *Biotechnol. J.* **2011**, *6*, 369–377. [CrossRef] [PubMed]

4. Mercier, S.M.; Diepenbroek, B.; Wijffels, R.H.; Streefland, M. Multivariate PAT solutions for biopharmaceutical cultivation: Current progress and limitations. *Trends Biotechnol.* **2014**, *32*, 329–336. [CrossRef] [PubMed]

5. Wold, S.; Sjöström, M.; Eriksson, L. PLS-regression: A basic tool of chemometrics. *Chemom. Intell. Lab. Syst.* **2001**, *58*, 109–130. [CrossRef]

6. Cruz, M.V.; Sarraguça, M.C.; Freitas, F.; Lopes, J.A.; Reis, M.A.M. Online monitoring of P(3HB) produced from used cooking oil with near-infrared spectroscopy. *J. Biotechnol.* **2015**, *194*, 1–9. [CrossRef] [PubMed]

7. Jain, G.; Jayaraman, G.; Kökpinar, Ö.; Rinas, U.; Hitzmann, B. On-line monitoring of recombinant bacterial cultures using multi-wavelength fluorescence spectroscopy. *Biochem. Eng. J.* **2011**, *58–59*, 133–139. [CrossRef]

8. Mazarevica, G.; Diewok, J.; Baena, J.R.; Rosenberg, E.; Lendl, B. On-line fermentation monitoring by mid-infrared spectroscopy. *Appl. Spectrosc.* **2004**, *58*, 804–810. [CrossRef] [PubMed]

9. Teixeira, A.P.; Portugal, C.A.M.; Carinhas, N.; Dias, J.M.L.; Crespo, J.P.; Alves, P.M.; Carrondo, M.J.T.; Oliveira, R. In situ 2D fluorometry and chemometric monitoring of mammalian cell cultures. *Biotechnol. Bioeng.* **2009**, *102*, 1098–1106. [CrossRef] [PubMed]

10. Ohadi, K.; Aghamohseni, H.; Legge, R.L.; Budman, H.M. Fluorescence-based soft sensor for at situ monitoring of chinese hamster ovary cell cultures. *Biotechnol. Bioeng.* **2014**, *111*, 1577–1586. [CrossRef] [PubMed]

11. Vázquez-Rey, M.; Lang, D.A. Aggregates in monoclonal antibody manufacturing processes. *Biotechnol. Bioeng.* **2011**, *108*, 1494–1508. [CrossRef] [PubMed]

12. Ring, J.; Seifert, J.; Jesch, F.; Brendel, W. Anaphylactoid reactions due to non-immune complex serum protein aggregates. *Monogr. Allergy* **1976**, *12*, 27–35.

13. Cromwell, M.E.; Hilario, E.; Jacobson, F. Protein aggregation and bioprocessing. *AAPS J.* **2006**, *8*, E572–E579. [CrossRef] [PubMed]

14. Zhang, Y.B.; Howitt, J.; McCorkle, S.; Lawrence, P.; Springer, K.; Freimuth, P. Protein aggregation during overexpression limited by peptide extensions with large net negative charge. *Protein Expr. Purif.* **2004**, *36*, 207–216. [CrossRef] [PubMed]

15. Gomez, N.; Subramanian, J.; Ouyang, J.; Nguyen, M.D.; Hutchinson, M.; Sharma, V.K.; Lin, A.A.; Yuk, I.H. Culture temperature modulates aggregation of recombinant antibody in CHO cells. *Biotechnol. Bioeng.* **2012**, *109*, 125–136. [CrossRef] [PubMed]

16. Paul, A.J.; Schwab, K.; Prokoph, N.; Haas, E.; Handrick, R.; Hesse, F. Fluorescence dye-based detection of mAb aggregates in CHO culture supernatants. *Anal. Bioanal. Chem.* **2015**, *407*, 1–8. [CrossRef] [PubMed]

17. Renard, D.; Lefebvre, J.; Griffin, M.; Griffin, W. Effects of pH and salt environment on the association of β-lactoglobulin revealed by intrinsic fluorescence studies. *Int. J. Biol. Macromol.* **1998**, *22*, 41–49. [CrossRef]

18. Ohadi, K.; Legge, R.L.; Budman, H.M. Intrinsic fluorescence-based at situ soft sensor for monitoring monoclonal antibody aggregation. *Biotechnol. Prog.* **2015**, *31*, 1423–1432. [CrossRef] [PubMed]

19. Hawe, A.; Sutter, M.; Jiskoot, W. Extrinsic Fluorescent Dyes as Tools for Protein Characterization. *Pharm. Res.* **2008**, *25*, 1487–1499. [CrossRef] [PubMed]

20. Khurana, R.; Coleman, C.; Ionescu-Zanetti, C.; Carter, S.A.; Krishna, V.; Grover, R.K.; Roy, R.; Singh, S. Mechanism of thioflavin T binding to amyloid fibrils. *J. Struct. Biol.* **2005**, *151*, 229–238. [CrossRef] [PubMed]

21. Jayaram, D.T.; Shankar, B.H.; Ramaiah, D. Effective Amyloid Defibrillation by Polyhydroxyl-Substituted Squaraine Dyes. *Chem. Asian J.* **2015**, *10*, 2689–2694. [CrossRef] [PubMed]

22. Capelle, M.A.; Gurny, R.; Arvinte, T. High throughput screening of protein formulation stability: Practical considerations. *Eur. J. Pharm. Biopharm.* **2007**, *65*, 131–148. [CrossRef] [PubMed]

23. Hawe, A.; Friess, W.; Sutter, M.; Jiskoot, W. Online fluorescent dye detection method for the characterization of immunoglobulin G aggregation by size exclusion chromatography and asymmetrical flow field flow fractionation. *Anal. Biochem.* **2008**, *378*, 115–122. [CrossRef] [PubMed]

24. Cummings, B.S.; Schnellmann, R.G. Measurement of cell death in mammalian cells. *Curr. Protoc. Pharmacol.* **2004**. [CrossRef]

25. Paul, A.J.; Schwab, K.; Hesse, F. Direct analysis of mAb aggregates in mammalian cell culture supernatant. *BMC Biotechnol.* **2014**, *14*, 99. [CrossRef] [PubMed]

26. Marose, S.; Lindemann, C.; Scheper, T. Two-Dimensional Fluorescence Spectroscopy: A New Tool for On-Line Bioprocess Monitoring. *Biotechnol. Prog.* **1998**, *14*, 63–74. [CrossRef] [PubMed]

27. Schwab, K.; Amann, T.; Schmid, J.; Handrick, R.; Hesse, F. Exploring the Capabilities of Fluorometric Online Monitoring on CHO Cell Cultivations Producing a Monoclonal Antibody. *Biotechnol. Prog.* **2016**. [CrossRef] [PubMed]

28. Schwab, K.; Lauber, J.; Hesse, F. Fluorometric In Situ Monitoring of an Escherichia coli Cell Factory with Cytosolic Expression of Human Glycosyltransferase GalNAcT2: Prospects and Limitations. *Bioengineering* **2016**, *3*, 32. [CrossRef]

29. Wise, B.M.; Gallagher, N.B.; Bro, R.; Shaver, J.M. *PLS Toolbox 3.0*; Eigenvector Research Inc.: Manson, WA, USA, 2003.

30. Roger, J.M.; Chauchard, F.; Bellon-Maurel, V. EPO-PLS external parameter orthogonalisation of PLS application to temperature-independent measurement of sugar content of intact fruits. *Chemom. Intell. Lab. Syst.* **2003**, *66*, 191–204. [CrossRef]

31. Chong, I.G.; Jun, C.H. Performance of some variable selection methods when multicollinearity is present. *Chemom. Intell. Lab. Syst.* **2005**, *78*, 103–112. [CrossRef]

32. He, F.; Phan, D.H.; Hogan, S.; Bailey, R.; Becker, G.W.; Narhi, L.O.; Razinkov, V.I. Detection of IgG aggregation by a high throughput method based on extrinsic fluorescence. *J. Pharm. Sci.* **2010**, *99*, 2598–2608. [CrossRef] [PubMed]

33. Amani, S.; Nasim, F.; Khan, T.A.; Fazili, N.A.; Furkan, M.; Bhat, I.A.; Khan, J.M.; Khan, R.H.; Naeem, A. Detergent induces the formation of IgG aggregates: A multi-methodological approach. *Spectrochim. Acta Part A Mol. Biomol. Spectrosc.* **2014**, *120*, 151–160. [CrossRef] [PubMed]

34. Franey, H.; Brych, S.R.; Kolvenbach, C.G.; Rajan, R.S. Increased aggregation propensity of IgG2 subclass over IgG1: Role of conformational changes and covalent character in isolated aggregates. *Protein Sci.* **2010**, *19*, 1601–1615. [CrossRef] [PubMed]

35. Hawe, A.; Filipe, V.; Jiskoot, W. Fluorescent molecular rotors as dyes to characterize polysorbate-containing IgG formulations. *Pharm. Res.* **2010**, *27*, 314–326. [CrossRef] [PubMed]

36. Bahram, M.; Bro, R.; Stedmon, C.; Afkhami, A. Handling of Rayleigh and Raman scatter for PARAFAC modeling of fluorescence data using interpolation. *J. Chemom.* **2006**, *20*, 99–105. [CrossRef]

bioengineering

MDPI

Article

Lagrangian Trajectories to Predict the Formation of Population Heterogeneity in Large-Scale Bioreactors

Maike Kuschel †, Flora Siebler † and Ralf Takors *

Institute of Biochemical Engineering, University of Stuttgart, 70569 Stuttgart, Germany;
Maike.Kuschel@ibvt.uni-stuttgart.de (M.K.); Flora.Siebler@ibvt.uni-stuttgart.de (F.S.)
* Correspondence: takors@ibvt.uni-stuttgart.de; Tel.: +49-711-685-64535
† These authors contributed equally to this work.

Academic Editor: Christoph Herwig
Received: 23 February 2017; Accepted: 24 March 2017; Published: 29 March 2017

Abstract: Successful scale-up of bioprocesses requires that laboratory-scale performance is equally achieved during large-scale production to meet economic constraints. In industry, heuristic approaches are often applied, making use of physical scale-up criteria that do not consider cellular needs or properties. As a consequence, large-scale productivities, conversion yields, or product purities are often deteriorated, which may prevent economic success. The occurrence of population heterogeneity in large-scale production may be the reason for underperformance. In this study, an *in silico* method to predict the formation of population heterogeneity by combining computational fluid dynamics (CFD) with a cell cycle model of *Pseudomonas putida* KT2440 was developed. The glucose gradient and flow field of a 54,000 L stirred tank reactor were generated with the Euler approach, and bacterial movement was simulated as Lagrange particles. The latter were statistically evaluated using a cell cycle model. Accordingly, 72% of all cells were found to switch between standard and multifork replication, and 10% were likely to undergo massive, transcriptional adaptations to respond to extracellular starving conditions. At the same time, 56% of all cells replicated very fast, with $\mu \geq 0.3$ h^{-1} performing multifork replication. The population showed very strong heterogeneity, as indicated by the observation that 52.9% showed higher than average adenosine triphosphate (ATP) maintenance demands (12.2%, up to 1.5 fold). These results underline the potential of CFD linked to structured cell cycle models for predicting large-scale heterogeneity *in silico* and *ab initio*.

Keywords: computational fluid dynamics; cell cycle model; Lagrange trajectory; scale-up; stirred tank reactor; population dynamics; energy level

1. Introduction

The physiological state of bacterial cells is strongly dependent on the surrounding conditions. As outlined in Müller et al. [1], external stress is a key factor inducing the formation of population heterogeneity, which differs according to growth phenotypes and cell cycle patterns. Moreover, concentration fluctuations occurring under large-scale mixing conditions have a measurable influence on growth and production yield [2–4]. Accordingly, homogeneity of the bacterial population may be affected, yielding subpopulations that co-exist next to each other [1]. Makinoshima et al. [5] observed five and ten cell populations of *Escherichia coli* during exponential growth and the subsequent stationary phase, respectively. For *Pseudomonas putida*, steady-state chemostat cultivation revealed that industry-like stress conditions induced changes in the cell cycle process. Under stress, deoxyribonucleic acid (DNA) replication was accelerated in a dose-dependent manner, yielding subpopulations with different DNA contents [6].

To investigate whether nutrient gradients of large-scale conditions foster the occurrence of population heterogeneity, the following concept was formulated. First, large-scale substrate gradients of a bioreactor should be simulated. Next, the path of bacterial cells through the gradients need to be tracked, and the resulting growth phenotypes monitored. Then, a cell cycle model can be used to translate changing growth conditions into cell cycle patterns. Apparently, this approach requires (i) a sound simulation of large-scale substrate gradients that trigger 'stress' in the cells and (ii) the translation of nutrient availability in growth patterns as a basis of cell cycle modelling. For the latter, the findings of Cooper and Helmstetter [7] were applied. They specified a relationship between chromosome content and cell cycle phase duration for *E. coli* B/r and showed that the amount of DNA varies continuously with the growth rate and substrate availability. Consequently, the durations of the cell cycle phases are strongly dependent on the environmental conditions.

The cell cycle of bacteria using binary fission can be divided into three parts: the time for initiation of replication (B-period), the time required for replication (C-period), and the time between replication and completed cell division (D-period). C-periods are the longest for slow-growing cells but decrease to constant values under elevated growth conditions [8]. In order to grow faster, replication and segregation are separated in time. Most bacteria initiate replication during a previous generation, leading to multifork replication.

Single-cell analysis by fluorescence-activated cell scanning has proven to be a valuable method to measure the DNA content from thousands of bacteria and to generate DNA content histograms for the population [9]. Also, latest lab on a chip techniques are a feasible method for measuring population heterogeneity [10,11]. Subpopulations with one, two, or more chromosomes can be detected. Skarstad [12] extended the model of Cooper and Helmstetter to calculate the number of individuals of *E. coli* B/r comprising a subpopulation with a specific DNA content from flow cytometry data. Furthermore, Skarstad determined the duration of the cell cycle periods for various growth rates. This was proven to be applicable for *P. putida* KT2440 as well [6].

It is still challenging to capture the magnitude and frequency of fluctuations in large scale bioprocesses and to predict the extent of the intracellular response. Several authors have suggested computational fluid dynamics (CFD) as a tool to provide detailed information of environmental conditions inside a fermenter. The gas, liquid, and bio phases are often modeled as a continuum by the Euler-Euler approach [13–15]. Typically, microorganisms react individually to different environmental conditions; therefore, a continuum description may not be advantageous. An extension of the Euler-Euler approach for the liquid phase is the use of population balance equations to model the heterogeneity of a population [16,17]. The incorporation of a detailed intracellular reaction network, however, demands a high computational effort to solve the complex distribution functions [18,19].

Since the pioneering work of Lapin et al. [20], environmental fluctuations have been studied from the perspective of microorganisms. The applied Euler-Lagrange approach uses a continuous representation of the fluid phase (Euler), combined with a segregated description of the cell population (Lagrange). The bacteria are simulated as particles, which are tracked on their way through the reactor. Statistical evaluation of these trajectories, denoted as bacterial lifelines, provide valuable information about substrate fluctuation frequencies experienced by microorganisms [21].

The influence of these fluctuations on cell cycle dynamics and energy levels has not been demonstrated yet. Thus, in this study, based on the work of Haringa et al. [21], an extensive statistical evaluation of bacterial lifelines was performed. Rather conservative operating conditions for the industrially relevant strain *P. putida* KT2440 were assumed to investigate the occurrence of and impact on population homogeneity. The Euler-Lagrange approach was combined with a cell cycle model of Lieder et al. [6] to gain deeper insights into the behaviors of cell cycle dynamics and individual distributions during large-scale fermentation.

These findings present a method to better analyze and understand the heterogeneity caused by scale up-induced stress stimuli.

2. Materials and Methods

2.1. Cell Cycle Model

Flow cytometry data ranging from $\mu = 0.1$ h^{-1} to 0.6 h^{-1} for *P. putida* KT2440 were obtained by Lieder et al. [6] and processed as shown in Figure 1. The data were channeled and displayed as the frequency distribution of DNA content. The durations of cell cycle phases C (DNA replication) and D (period between replication and completed cell division) were determined iteratively by minimizing the deviation between experimental and theoretical DNA histograms. The theoretical DNA content of an asynchronous, ideal culture in which all cells have equal growth parameters was derived from the age distribution according to Cooper and Helmstetter [7]. Using this probability density function for cells of a specific cell age, Cooper and Helmstetter further calculated the theoretical chromosome content per cell at a specific cell age. This model was extended by Skarstad et al. [12] to calculate the frequency of the occurrence of a specific DNA content in an interval of ongoing DNA synthesis. The durations of phases C and D are decisive for the distribution of DNA content. Different values for C were obtained to fit the experimental histograms for various growth rates. Based on the work of Lieder [22], a function for C-phase duration, dependent on the growth rate of *P. putida* KT2440, was derived. A correlation for C proposed by Keasling et al. [23] was used.

$$C = C_{min} \left(1 + a\, e^{b\,\mu}\right) \tag{1}$$

where C is the length of the C period, C_{min} is the minimal length of the C period, μ represents the growth rate and a and b are parameters that fit the experimental data. Based on the experimental data of Lieder et al. [6], the parameter estimation resulted in $C_{min} = 0.77$ h, a = 1.83, and b = 4.88.

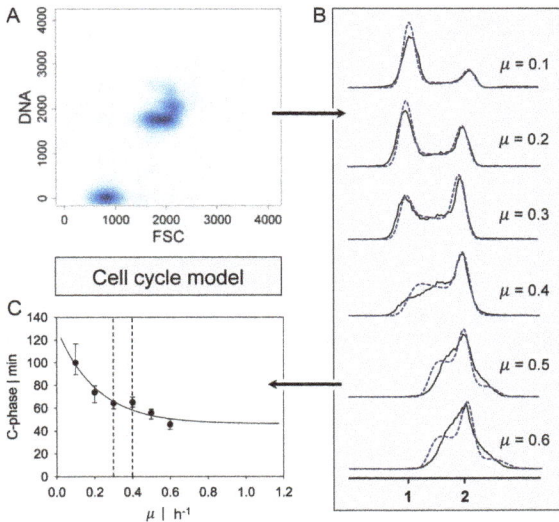

Figure 1. Approach for the cell cycle dynamics model. (**A**) Representative flow cytometry scatter plot for deoxyribonucleic acid (DNA) content of the growth rate $\mu = 0.3$ h^{-1}. (**B**) DNA content over counts for growth rates ranging from $\mu = 0.1$ h^{-1} up to $\mu = 0.6$ h^{-1}. A single genome is indicated by 1, and double chromosomes by 2. Black lines present experimental data, and blue dashed lines present the calculation of the cell cycle model. (**C**) Approximated C-phase duration over growth rate estimated by the cell cycle model (1% parameter covariance). Black dashed lines indicate the transition regime from single-forked to multiforked replication. Flow cytometry data obtained by Lieder et al. [6].

2.2. Numerical Simulation

2.2.1. Geometry and Reactor Setup

In order to generate a pseudostationary glucose gradient of an industrial fed batch fermentation, a large-scale stirred tank bioreactor was chosen. Precise dimensions and information about the inner geometry can be found in Appendix A (Figure A1 and Table A1). The main geometry was derived from Haringa et al. [21] and only slightly modified for the purpose of this study. With an H/D ratio of 2.57, the total volume was about 54,000 L. The reactor setup included four baffles and a stirrer with two Rushton agitators. The lower stirring unit was equipped with eight blades, and the middle unit with six blades. With a stirring rate of 100 rpm, a tip speed of 5–8 m s^{-1} was reached. The impeller Reynolds number was 1.8×10^6, the power number 13.15, and the needed power was 226 kW.

The feeding rate was set as half of the maximum uptake rate $q_{s,max}$ of *P. putida* with 0.738 kg$_{glc}$·kg$_{CDW}^{-1}$·h^{-1}. Aeration, gas transfer, and oxygen uptake were neglected in this study. Therefore, no gassing system was installed. A cell concentration of 10 kg$_{CDW}$·m^{-3} was assumed, and a simple Monod-like kinetic was used to simulate the substrate uptake q_s:

$$q_s = q_{s,max} \cdot \frac{c_s}{K_s + c_s} \tag{2}$$

where $q_{s,max}$ is the maximum uptake rate, c_s is the glucose concentration, and the approximated substrate specific uptake constant K_s with 10 mg·L^{-1}. The maximum uptake rate was calculated with the biomass substrate yield $Y_{XS} = 0.40$ g$_s$·g$_{CDW}^{-1}$ and the maximum growth rate $\mu = 0.59$ h^{-1} [22,24].

2.2.2. Simulation Setup

For the numerical simulation, the commercial calculation tool ANSYS Fluent version 17.0 was used. Using this finite volume-based fluid dynamic analysis program, the virtual geometry was built, and spatial discretization was performed. A total of 445,000 numerical cells yielded the same circulation time as achieved by Haringa et al. [21]. The flow field was approximated by solving the Reynolds-averaged Navier-Stokes (RANS) equations in combination with the standard k-ε model for turbulence. All surfaces were set as slip boundaries, except for the frictionless top area, which implied the reactor filling height. Both impeller units were set to sliding mesh motion to generate a more realistic flow field.

For glucose feed, a separate volume at the top of the reactor was defined, and a constant mass flow was set. The feed was inserted as mass percentage, with constant pressure and volume. The hydrodynamic and kinetic was calculated every 10 ms until the overall glucose concentration was constant and a pseudostationary gradient was reached. Finally, an average flow field and glucose gradient were obtained over 150 s. In further simulations, the hydrodynamic and glucose gradient were set as frozen.

Bacteria lifelines were simulated as massless Lagrangian particles with a discrete random walk (DRW) model passing through the flow field. Every 30 ms, the position and glucose concentration for each bacterium were recorded. In total, 120,000 bacterial cells were tracked over 260 s. According to the ergodic theorem, the same average values are obtained by tracking 1,560,000 bacteria for 20 s (the approximate circulation time). The simulation would yield even more precise statistical evaluations by increasing the number of lifelines.

2.3. Statistical Evaluation

All bacterial lifelines were evaluated statistically and grouped according to the regime borders. The growth rate was calculated for each bacterial cell and each time interval. The regimes were classified as follows: standard forked replication S for $\mu \leq 0.3$ h^{-1}, the transition area T ($0.3 < \mu < 0.4$ h^{-1}), and multifork replication M for $\mu \geq 0.4$ h^{-1} derived by the cell cycle model (see Section 2.1.). By evaluating the cell history, further classifications were made. Six regime transitions follow when two transitions and one retention time were considered:

- **STM**: transition from standard forked to multiforked with a retention time in the transition area.
- **STS**: standard forked, retention in the transition area, and back to standard forked
- **TST**: starting from the transition area with retention in a single forked area and back to transition
- **MTS**: multiforked replication regime to single forked replication with a retention time in the transition area
- **MTM**: beginning in the multifork regime with retention in the transition area and back to the multifork regime
- **TMT**: circulation from transition back to transition area with retention time in the multifork replication regime

The second capital letter always indicates the area in which the retention time τ was measured. Before the bacterial lifelines were grouped in regimes, a moving-average filter was applied to filter unrealistic, turbulent fluctuations caused by the standard DRW model (see Appendix B). A second one-dimensional (1D) filter was conducted to erase rapid sequential regime transitions smaller than 0.09 s. Both filtering steps caused deviations from the raw data of less than 5%.

The distribution of the growth rates was derived by calculating the mean growth rate for the whole reactor and the mean growth rate for 20 s for each bacterium. This distribution combined with the cell cycle approach resulted in a distribution of different C-phase durations using Equation (1). Additionally, the energy level distribution was obtained based on Pirt's law [25]:

$$q_{ATP} = \frac{\mu}{Y_{x/ATP}} + m_{ATP} \tag{3}$$

with the *Pseudomonas putida* properties of nongrowth-associated maintenance $m_{ATP} = 3.96 \text{ mmol}_{ATP} \cdot g_{CDW}^{-1} \cdot h^{-1}$ and the growth-associated maintenance $Y_{XATP} = \frac{1}{85} g_{CDW} \cdot \text{mmol}_{ATP}^{-1}$ [24].

3. Results and Discussion

In order to investigate heterogeneity in large-scale bioreactors, a pseudostationary glucose gradient occurring during fed batch fermentation of *P. putida* was simulated. Therefore, a biomass of 10 kg·m^{-3} was assumed, which remained constant within the time observed. For higher biomass concentrations, stronger gradients can be expected.

3.1. Gradient and Flow Field

In a 54,000 L stirred tank reactor, a pseudostationary glucose gradient was obtained with CFD simulations. The average glucose concentration was monitored until no further changes could be observed. The residual steady state glucose concentration was 20.7 mg·L^{-1}. The theoretical growth rate for every numerical cell was computed (Eulerian approach), resulting in an average growth rate of $\mu = 0.294 \text{ h}^{-1}$. Ideal mixing was assured by comparing the average growth rate in the reactor (Eulerian approach) and the expected growth rate for the set feed rate $\mu = 0.295 \text{ h}^{-1}$. In the fed batch fermentation, the feeding rate amounted to half the maximum uptake rate of *P. putida*. The objective of the simulation was to generate a realistic glucose gradient with concentrations for which theoretical growth rates ranging from 0.0 h^{-1} to 0.59 h^{-1} could be approximated. Moreover, the distribution of bacteria that were introduced from different vertical positions in the reactor at the start of the simulation is displayed.

In Figure 2, three reactor cross sections are depicted to describe (A) the growth rate regimes (see also Section 2.3), (B) the flow field, and (C) the bacterial distribution. Due to asymmetric reactor geometry (see Section 2.2.1), the mean flow field and mean glucose gradient showed periodic changes. Accordingly, the averages of the flow field and gradients over 150 s were computed to track the bacteria (Figure 2C) as lifelines. Bacteria moved faster when approaching the stirrer. This clearly indicated zones with different residence times. However, tracking the bacterial paths showed that they evenly crossed every part in the reactor.

The underlying gradient was not expected to perfectly reflect the real experiment. Several assumptions had to be made. For simplicity, bubbling flow and oxygen transfer were neglected. The kinetic reaction of substrate consumption following a Monod-like kinetic was assumed to take place in every numerical cell. This implied that the bacterial cells were distributed homogeneously at each time step, which is only a simplified scenario (Figure 2C). However, to examine the effects of cell history or lag phases of the bacteria on the gradient itself, an existing gradient had to be installed with the stated simplifications. In the following sections, a detailed statistical analysis is provided to study the influence of the gradient on the bacteria and reverse in a realistic manner.

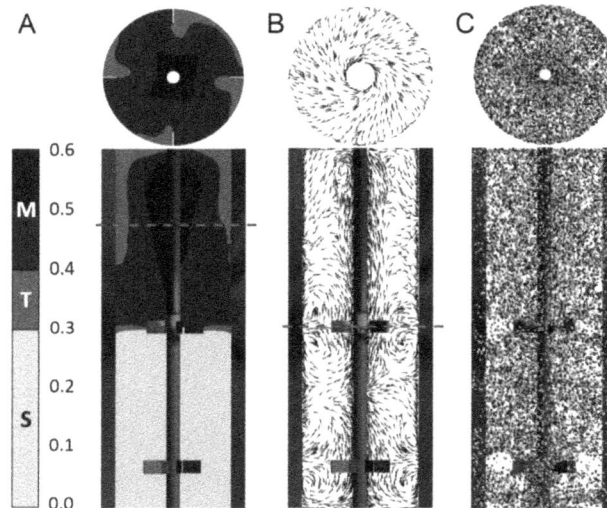

Figure 2. Simulation of gradients and bacterial lifelines. (**A**) Averaged substrate gradient calculated for 150 s, colored by regime classification: standard replication S ($\mu < 0.3$) in light gray, transition regime T ($0.3 \leq \mu \leq 0.4$) in gray, and multifork replication M ($\mu > 0.4$) in dark gray. (**B**) Average flow field estimated for 150 s. (**C**) Representative magnified bacteria particles (around 2000) at a certain time step (colored by particle ID; low numbers in dark gray represent a starting point close to the reactor bottom, high numbers in light gray represent a starting point close to the reactor top). Horizontal section planes are indicated by dashed red lines; otherwise, the top view is shown.

3.2. Lagrangian Trajectory

For 260 s, 120,000 bacteria were tracked on their paths crossing different substrate concentrations. Figure 3 depicts growth rate profiles of two organisms for 20 s, referred to as lifelines L1 and L2. Figure 3C shows the related paths.

According to the regime thresholds (see Section 2.3 and Figure 3A, dashed lines), the growth rate trajectories could be transferred to replication modus curves, as described in Figure 3B). The lifeline L1 revealed high variations in glucose concentrations that were likely to induce strong metabolic changes. In contrast, environmental shifts along L2 were moderate, and there were no effects on metabolism or the cell cycle. The first lifeline L1 gave information regarding five regime transition strategies (STS, TST, STM, TMT, and MTS) and the individual residence times. Lifelines L1 and L2 started from different positions in the reactor and were unequal in length because they moved according to the predominant velocity field. Within 20 s, L2 did not approach the feed zone, remaining in an area of reduced substrate concentration and increased shear stress, owing to the higher velocity of L2.

As shown in Figure 3B,C, within a defined timescale, bacteria completely sensed different environmental conditions. Whereas L2 seemed to remain in the same environment, L1 passed different glucose concentrations and performed several replication strategies. Each metabolic adjustment will cost energy and could have an impact on the production yield.

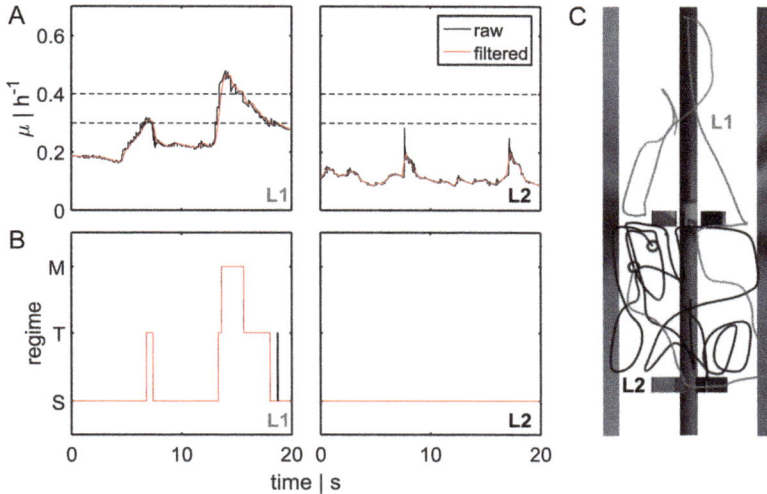

Figure 3. Bacterial lifeline and regime transition classification. (**A**) Two-dimensional (2D) bacterial lifeline for different growth rates μ over time. The black line represents raw data, and the red line represents filtered data (moving average filter to correct discrete random walk (DRW) fluctuations). Black dashed lines indicate the transition regime from single-forked to multiforked replication. (**B**) Translation of filtered (one-dimensional (1D) filter) growth rate curves for the three regimes: multifork replication regime M, transition between standard forked and multiforked T, and standard replication S. Examples for two bacterial lifelines L1 and L2 are depicted. For L1, five regime transitions (STS, TST, STM, TMT, and MTS; see Section 2.3) were analyzed. (**C**) Bacterial movement patterns for two bacterial lifelines (L1 in gray and L2 in black). Starting positions are indicated by black circles.

3.3. Statistical Evaluation

3.3.1. Regime Transition Frequency

All bacterial lifelines were scanned for regime transitions and retention times in order to obtain the frequency distributions as a function of τ. Thus, six transition strategies were evaluated in a statistical manner to gain insights into cell histories and possible cell behaviors (see also Section 2.3).

Figure 4 shows the counts for each regime transition at a certain retention time. All regime transition statistics, except the TST transition, exhibited a decay after at least 10 s. Bacteria starting from the transition regime T could remain in an area of low concentration for up to 73.5 s (data not shown), where they could grow regularly (standard forked S), before changing back to the T regime. This could be explained by the flow field and gradient pictured in Figure 2A,B. The critical concentrations representing possible growth rates for the regime transition ($\mu \geq 0.3$ h^{-1} and $\mu > 0.4$ h^{-1}) were located in the upper half of the reactor. Rushton turbines usually cause flow patterns moving away from the blades to the wall, where they circulate up or down, thereby forming large eddies for each stirrer set (Figure 2B). Consequently, cells will often circulate in this segment and do not pass other areas of the reactor. The lower part of the reactor, which does not provoke a regime transition and, therefore, badly supplies the organisms with substrate, consisted of three segments. As a result, the average retention time in the TST transition was the longest ($\overline{\tau}_{TST}$ = 8.54 s). All other average and maximum retention times are listed in Table 1. The shapes of the distributions follow a Poisson distribution. The maximal retention time was defined as the limit, within which 99% of the values were located.

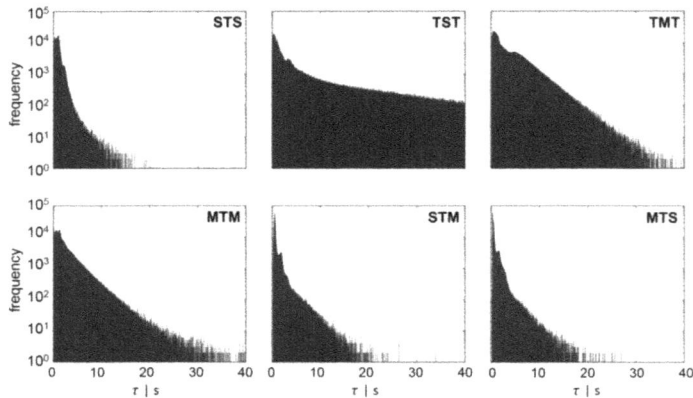

Figure 4. Regime transition frequency as a function of the retention time τ. Regime transition classifications are indicated in the left corner of each panel. The second capital letter always indicates the area, in which the retention time τ was measured. The regime transition count for each retention time was scaled logarithmically.

Table 1. Average and maximal retention time in a specific regime. For the six regimes (STS, TST, TMT, MTM, STM, and MTS), the average ($\overline{\tau}$) and maximal retention times (τ_{max}) are displayed in seconds. The maximum τ was defined as the limit, within which 99% of the values were located.

Regime Transition	$\overline{\tau}$ [s]	τ_{max} [s]
STS	0.99	3.7
TST	8.54	73.5
TMT	3.53	16.25
MTM	2.45	13
STM	0.95	6.6
MTS	0.88	5.5

Lifeline statistics provide insights into the frequency of regime transitions and residence times. Depending on the cell history, i.e., the concentrations of bacteria encountered before the bacteria passed the actual concentration, the cells will adapt accordingly. Although metabolic adaptation is known to be very rapid, the initiation of regulatory programs involving transcriptional changes is slower. Investigating the impact of large-scale conditions for *E. coli*, Löffler et al. [26] showed that fundamental transcriptional programs were initiated after 70 s of glucose shortage. After 30 s, metabolic consequences were measured, and the first transcriptional changes were detected. In total, about 600 genes were found to be up- or downregulated repeatedly, indicating a strong adaption.

Considering this finding during the regime analysis, it is assumed that all cells travelling from high (M) to low (S) substrate availability should be influenced. Being prepared for multifork replication in M, the cells must adapt to standard replication (S). By analogy, this also includes travelers from T to S. Such cells can have a growth rate of about 0.4 h^{-1} before they adapt to growth rates of less than 0.3 h^{-1}. During the observation window of 260 s, 72.6% of all cells were expected to carry out this move at least once and to linger more than 30 s in regime S. About 14.7% of all cells were expected to stay more than 70 s in regime S after experiencing higher glucose concentrations in regime T. Furthermore, if a regime transition from maximal to moderate growth conditions (MTS) with the retention time in regime T and S is assumed, 55.5% of all cells performed this move for more than 30 s. A retention time of 70 s was calculated for 10.4% of all cells. The time scales of 30 s and 70 s were shown to significantly influence the transcriptional response of *E. coli* [26], leading to the assumption that changes in adenosine triphosphate (ATP) and guanosine triphosphate (GTP) levels of *P. putida* KT2440 could also be expected.

3.3.2. Energy and C-Phase Duration Distribution

For the observation window of 260 s, the growth rate profiles of 120,000 bacteria were calculated. Given the set feed rate, the average μ of 0.295 h^{-1} was expected. Using the Lagrangian approach, an average growth rate of $\mu = 0.269$ h^{-1} was computed, indicating an adequate deviation of 8.5% compared to the Eulerian approach with $\mu = 0.294$ h^{-1} (see Section 3.1).

The distribution of the ATP consumption rate q_{ATP} is presented in Figure 5A. The growth rate μ and q_{ATP} were not evenly distributed compared to the mean value, but exhibited individual distributions according to the gradient. The ATP consumption rate was calculated applying Pirt's law (see Equation (3)). While only 6.3% of all cells had a mean ATP consumption rate of $q_{ATP,mean} = 29.31 \pm 2$ mmol$_{ATP} \cdot g_{CDW}^{-1} \cdot h^{-1}$, 40.8% showed a reduced consumption rate of less than 27.31 mmol$_{ATP} \cdot g_{CDW}^{-1} \cdot h^{-1}$, and 52.9% showed an increased energy demand of 31.31 mmol$_{ATP} \cdot g_{CDW}^{-1} \cdot h^{-1}$ in comparison to the average consumption rate. Moreover, 12.2% show an energy demand that was more than 1.5 times that of the mean value in the reactor.

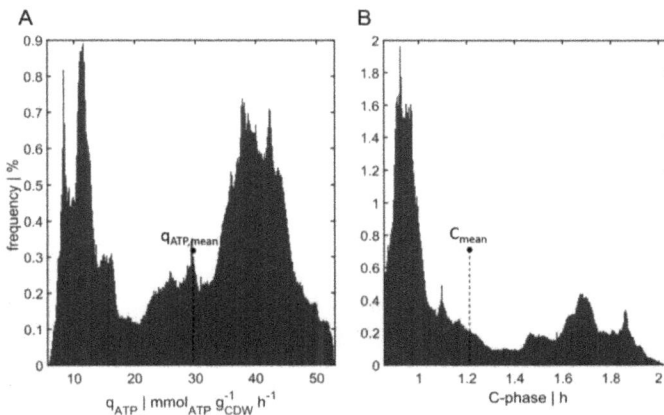

Figure 5. Distribution of C-phase duration and energy level. (**A**) Frequencies of cells with a specific adenosine triphosphate (ATP) consumption rate (q_{ATP}) tracked for 20 s. Average value of $q_{ATP,mean} = 29.31$ mmol$_{ATP} \cdot g_{CDW}^{-1} \cdot h^{-1}$. Range of the x-axis from $q_{ATP,min} = 5.57$ mmol$_{ATP} \cdot g_{CDW}^{-1} \cdot h^{-1}$ to $q_{ATP,max} = 52.98$ mmol$_{ATP} \cdot g_{CDW}^{-1} \cdot h^{-1}$. (**B**) Frequency of cells having a specific duration of replication (C-phase). Average C-phase duration of $C_{mean} = 1.21$ h. Range of the x-axis from $C_{min} = 0.86$ h to $C_{max} = 2.05$ h. Counts were divided into 300 bins.

The distribution will differ if increased nongrowth-associated maintenance m_{ATP} is considered. As outlined by Löffler et al. [26], m_{ATP} increases by 40–50% when cells are exposed to large-scale substrate gradients.

The individual growth profiles of the cells are the basis for deducing cell cycle patterns using the cell cycle model (see Section 2.3). Distributions of the C-length (encoding DNA replication) could be derived for the population of 120,000 bacteria. Figure 5B shows the average duration of replication of 1.21 h and the frequency of cells with a C-phase duration ranging from $C_{min} = 0.86$ h to $C_{max} = 2.05$ h. Clearly, the bacteria were not evenly distributed according to the mean value, and there was a large heterogeneity in the reactor. Although only 22.3% of all cells had a replication phase of 1.21 ± 0.2 h, about 30% possessed a C-period of more than 1.41 h. In contrast, 47.7% displayed a shorter replication phase than the average time for replication (less than 1.01 h). Moreover, approximately 56.1% of the cells were rapidly replicating cells with a growth rate higher than $\mu = 0.3$ h^{-1}. For these cells, it can be assumed that they already started to completely adjust their metabolism to achieve multifork replication. As shown in Figure 5B, the bioreactor population was strongly heterogeneous, characterized by a nonequal distribution of bacteria in different cell cycle states. Three different growth phenotypes are shown: C-phase durations of (i) 0.94 ± 0.08 h, (ii) 1.68 ± 0.1 h, and (iii) a transition state of C-phases ranging from 1.1 to 1.5 h. Previously, subpopulations resulting from chemostat experiments have been categorized in populations containing one, two, or more than two chromosomes [27]. With this simulation setup, a model-based superposition of subpopulations containing different growth rates to mimic the scenario in a (fed)batch fermentation was shown. For the underlying gradient, new categories of subpopulations according to the C-phase durations mentioned above can be formulated.

4. Conclusions

The existence of population heterogeneity in industrial fermenters has been demonstrated, but it still not completely understood. Improvements in fermenter operation, reactor design, and strain engineering can be achieved as more information of cell behaviors during large-scale production becomes available. In this study, the formation of heterogeneity by combining CFD with a cell cycle model of *P. putida* was investigated. With this method, heterogeneity can be interpreted from the bacterial point of view, particularly with respect to the growth phase durations and energy demands of the cell.

Average and maximum residence times for each transition strategy have been approximated and can be linked to scale-down experiments using STR-PFR setups. Moreover, distributions of growth rates, ATP consumptions, and C-phase durations could be generated. Such findings provide important insights into the intracellular mechanisms that determine growth phenotypes. These mechanisms may become a crucial part of strain and process engineering to predict *ab initio* and *in silico* whether and how large-scale performance will meet expectations. Realistic large-scale cultivation can be simulated by investigating the "subpopulations" individually. Specifically, it may be possible to elucidate whether the total drop in production performance during large-scale production is caused by all cells or by individual "subpopulations" that underperform.

To further investigate such problems, heterogeneity studies need to be coupled with single-cell product kinetics. Moreover, research will need to focus on the quantitative measurement of the impact of stress intensity on the m_{ATP} level. This will enable prediction of the total energy demand for a given setup.

Acknowledgments: The research could not have been conducted without the previous work on the cell cycle model of Sarah Lieder and the flow cytometry analysis of Susann Müller. We also thank Alexei Lapin and Alexander Nieß for their advice and support during this study. This work was supported by the German Research Foundation (DFG; GZ: TA241/5-2) and the Federal Ministry of Education and Research (BMBF; FKZ031A468B).

Author Contributions: Ralf Takors initiated the study and provided important advice during the entire investigation. Maike Kuschel and Flora Siebler equally conducted all parts of the simulation, evaluation, and writing of the manuscript as a team.

Conflicts of Interest: The authors declare no conflicts of interest.

Appendix A

More precise information of the reactor setup and geometry can be found in Table A1 and Figure A1.

Table A1. Dimensions of the reactor setup pictured in Figure A1.

Description	Symbol	Relation
Reactor diameter	D_R	3.00 m
Impeller diameter	D_I	0.43 D_R
Impeller height	H_I	0.21 D_I
Bottom clearance	C_1	0.30 D_R
Impeller spacing	ΔC	1.00 D_R
Upper clearance	C_2	1.27 D_R
Baffle width	B	0.10 D_R
Liquid height	H_L	$C_1 + \Delta C + C_2$

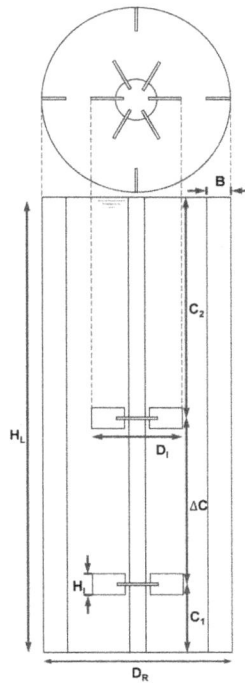

Figure A1. Schematic diagram of reactor geometry derived from Haringa et al. [21]. The stirred tank reactor contains four baffles and two Rushton turbines with eight blades (**bottom**) and six blades (**top**). Dimensions indicated by capital letters are explained in Table A1.

Appendix B

The standard moving average filter of MATLAB is a linear filter (low pass filter), which removes high frequency components such as fluctuations caused by the DRW model. It is formulated as:

$$m(t) = \sum_{j=-q}^{q} y_{t+j} \quad q < t < N - q \tag{A1}$$

with:

$$q = \frac{\bar{\tau} - 1}{2} \tag{A2}$$

where N is the total number of measured time points and $\bar{\tau}$ the filter timescale.

References

1. Müller, S.; Harms, H.; Bley, T. Origin and analysis of microbial population heterogeneity in bioprocesses. *Curr. Opin. Biotechnol.* **2010**, *21*, 100–113. [CrossRef] [PubMed]
2. Bylund, F.; Collet, E.; Enfors, S.; Larsson, G. Substrate gradient formation in the large-scale bioreactor lowers cell yield and increases by-product formation. *Bioprocess Eng.* **1998**, *18*, 171–180. [CrossRef]
3. Enfors, S.; Jahic, M.; Rozkov, A.; Xu, B.; Hecker, M.; Ju, B. Physiological responses to mixing in large scale bioreactors. *J. Biotechnol.* **2001**, *85*, 175–185. [CrossRef]
4. Takors, R. Scale-up of microbial processes: Impacts, tools and open questions. *J. Biotechnol.* **2012**, *160*, 3–9. [CrossRef] [PubMed]
5. Makinoshima, H.; Nishimura, A.; Ishihama, A. Fractionation of *Escherichia coli* cell populations at different stages during growth transition to stationary phase. *Mol. Microbiol.* **2002**, *43*, 269–279. [CrossRef] [PubMed]
6. Lieder, S.; Jahn, M.; Koepff, J.; Müller, S.; Takors, R. Environmental stress speeds up DNA replication in *Pseudomonas putida* in chemostat cultivations. *Biotechnol. J.* **2016**, *11*, 155–163. [CrossRef] [PubMed]
7. Cooper, S.; Helmstetter, C.E. Chromosome replication and the division cycle of *Escherichia coli*. *J. Mol. Biol.* **1968**, *31*, 519–540. [CrossRef]
8. Girault, M.; Kim, H.; Arakawa, H.; Matsuura, K.; Odaka, M.; Hattori, A.; Terazono, H.; Yasuda, K. An on-chip imaging droplet-sorting system: A real-time shape recognition method to screen target cells in droplets with single cell resolution. *Sci. Rep.* **2017**, *7*, 1–10. [CrossRef] [PubMed]
9. Cheng, Y.H.; Chen, Y.C.; Brien, R.; Yoon, E. Scaling and automation of a high-throughput single-cell-derived tumor sphere assay chip. *Lab Chip* **2016**, *16*, 3708–3717. [CrossRef] [PubMed]
10. Helmstetter, C.E. Timing of Synthetic Activities in the Cell Cycle. In *Escherichia coli and Salmonella. Cellular and Molecular Biology*; Neidhardt, F.C., Ed.; American Society for Microbiology (ASM) Press: Washington, DC, USA, 1996; pp. 1627–1639.
11. Müller, S. Modes of cytometric bacterial DNA pattern: A tool for pursuing growth. *Cell Prolif.* **2007**, *40*, 621–639. [CrossRef] [PubMed]
12. Skarstad, K.; Steen, H.B.; Boye, E. *Escherichia coli* DNA distributions measured by flow cytometry and compared with theoretical computer simulations. *J. Bacteriol.* **1985**, *163*, 661–668. [PubMed]
13. Larsson, G.; Törnkvist, M.; Ståhl Wernersson, E.; Trägårdh, C.; Noorman, H.; Enfors, S.O. Substrate gradients in bioreactors: Origin and consequences. *Bioprocess Eng.* **1996**, *14*, 281–289. [CrossRef]
14. Noorman, H.; Morud, K.; Hjertager, B.H.; Traegaardh, C.; Larsson, G.; Enfors, S.O. CFD modeling and verification of flow and conversion in a 1 m^3 bioreactor. *BHR Gr. Conf. Ser. Publ.* **1993**, *5*, 241–258.
15. Schmalzriedt, S.; Jenne, M.; Mauch, K.; Reuss, M. Integration of physiology and fluid dynamics. *Adv. Biochem. Eng.* **2003**, *80*, 19–68.
16. Morchain, J.; Gabelle, J.-C.; Cockx, A. A coupled population balance model and CFD approach for the simulation of mixing issues in lab-scale and industrial cioreactors. *Am. Inst. Chem. Eng.* **2014**, *60*, 27–40. [CrossRef]
17. Bezzo, F.; Macchietto, S.; Pantelides, C.C. General hybrid multizonal/CFD approach for bioreactor modeling. *AIChE J.* **2003**, *49*, 2133–2148. [CrossRef]
18. Mantzaris, N.V.; Liou, J.; Daoutidis, P.; Srienc, F. Numerical solution of a mass structured cell population balance model in an environment of changing substrate concentration. *J. Biotechnol.* **1999**, *71*, 157–174.
19. Henson, M.A. Dynamic modeling of microbial cell populations. *Curr. Opin. Biotechnol.* **2003**, *14*, 460–467. [CrossRef]
20. Lapin, A.; Müller, D.; Reuss, M. Dynamic behavior of microbial populations in stirred bioreactors simulated with Euler-Lagrange methods: Traveling along the lifelines of single cells. *Ind. Eng. Chem. Res.* **2004**, *43*, 4647–4656. [CrossRef]
21. Haringa, C.; Tang, W.; Deshmukh, A.T.; Xia, J.; Reuss, M.; Heijnen, J.J.; Mudde, R.F.; Noorman, H.J. Euler-Lagrange computational fluid dynamics for (bio)reactor scale-down: An analysis of organism life-lines. *Eng. Life Sci.* **2016**, *16*, 652–663. [CrossRef] [PubMed]
22. Lieder, S. *Deciphering Population Dynamics as a Key for Process Optimization*; University of Stuttgart: Stuttgart, Germany, 2014.
23. Keasling, J.D.; Kuo, H.; Vahanian, G. A Monte Carlo simulation of the *Escherichia coli* cell cycle. *J. Theor. Biol.* **1995**, *176*, 411–30. [CrossRef] [PubMed]

24. Van Duuren, J.B.J.H.; Puchałka, J.; Mars, A.E.; Bücker, R.; Eggink, G.; Wittmann, C.; Dos Santos, V.A.P.M. Reconciling in vivo and *in silico* key biological parameters of *Pseudomonas putida* KT2440 during growth on glucose under carbon-limited condition. *BMC Biotechnol.* **2013**, *13*, 93. [CrossRef] [PubMed]
25. Pirt, S.J. The maintenance energy of bacteria in growing cultures. *Proc. R. Soc. Lond. Ser. B. Biol. Sci.* **1965**, *163*, 224–231. [CrossRef]
26. Löffler, M.; Simen, J.D.; Jäger, G.; Schäferhoff, K.; Freund, A.; Takors, R. Engineering *E. coli* for large-scale production—Strategies considering ATP expenses and transcriptional responses. *Metab. Eng.* **2016**, *38*, 73–85. [CrossRef] [PubMed]
27. Lieder, S.; Jahn, M.; Seifert, J.; von Bergen, M.; Müller, S.; Takors, R. Subpopulation-proteomics reveal growth rate, but not cell cycling, as a major impact on protein composition in *Pseudomonas putida* KT2440. *AMB Express* **2014**, *4*, 71. [CrossRef] [PubMed]

bioengineering

MDPI

Article

Hypoxic Three-Dimensional Scaffold-Free Aggregate Cultivation of Mesenchymal Stem Cells in a Stirred Tank Reactor

Dominik Egger [1,†], Ivo Schwedhelm [2,†], Jan Hansmann [2] and Cornelia Kasper [1,*]

1   Department of Biotechnology, University of Natural Resources and Life Sciences,
     Muthgasse 18, 1190 Vienna, Austria; dominik.egger@boku.ac.at
2   Translational Center, University Hospital Wuerzburg, Roentgenring 11, 97070 Wuerzburg, Germany;
     ivo.schwedhelm@uni-wuerzburg.de (I.S.); jan.hansmann@uni-wuerzburg.de (J.H.)
*   Correspondence: cornelia.kasper@boku.ac.at; Tel.: +43-1-47654-79030
†   These authors contributed equally to this work.

Academic Editor: Christoph Herwig
Received: 26 April 2017; Accepted: 21 May 2017; Published: 23 May 2017

Abstract: Extensive expansion of mesenchymal stem cells (MSCs) for cell-based therapies remains challenging since long-term cultivation and excessive passaging in two-dimensional conditions result in a loss of essential stem cell properties. Indeed, low survival rate of cells, alteration of surface marker profiles, and reduced differentiation capacity are observed after in vitro expansion and reduce therapeutic success in clinical studies. Remarkably, cultivation of MSCs in three-dimensional aggregates preserve stem cell properties. Hence, the large scale formation and cultivation of MSC aggregates is highly desirable. Besides other effects, MSCs cultivated under hypoxic conditions are known to display increased proliferation and genetic stability. Therefore, in this study we demonstrate cultivation of adipose derived human MSC aggregates in a stirred tank reactor under hypoxic conditions. Although aggregates were exposed to comparatively high average shear stress of 0.2 Pa as estimated by computational fluid dynamics, MSCs displayed a viability of 78–86% and maintained their surface marker profile and differentiation potential after cultivation. We postulate that cultivation of 3D MSC aggregates in stirred tank reactors is valuable for large-scale production of MSCs or their secreted compounds after further optimization of cultivation parameters.

Keywords: mesenchymal stem cells; scaffold-free; aggregate cultivation; stirred tank reactor; dynamic cultivation; hypoxia; stemness; computational fluid dynamics

1. Introduction

In the context of regenerative medicine, mesenchymal stem cells (MSCs) are still considered the most promising and eligible candidate for therapeutic use in cell-based therapies. Their regenerative potential is based on high proliferative activity, the capacity to differentiate into specific cell types of the musculoskeletal and connective tissue [1,2] as well their ability to specifically migrate to injured tissue sites, where they are involved in tissue repair and anti-inflammatory effects [3] via delivery of trophic factors [4–6]. Consequently, this results in immunosuppressive effects, enhanced tissue repair, and angiogenesis. Therefore, it is essential to maintain their inherent properties during ex vivo cultivation to enable for therapeutic success and reproducibility in clinical studies.

However, altered immune properties and low in vivo survival rates of MSCs were reported after ex vivo expansion [7,8]. Still, large-scale expansion of MSCs is usually carried out in two-dimensional (2D) static conditions, which was shown to alter their inherent immunophenotype [9]. In contrast, the formation of three-dimensional (3D) MSC aggregates seems to preserve their phenotype and differentiation potential [10,11]. Furthermore, increased secretion of proangiogenic factors and anti-inflammatory cytokines was observed after aggregate formation [12,13]. Therefore, the upscale of aggregate formation and cultivation for the therapeutic use of cells or their secreted compounds is desirable. The cultivation of aggregates in small-scale systems like microtiter plates or hanging drops is well established [14]. However, aggregates display nutrient and oxygen gradients from surface to core which was demonstrated to result in a necrotic core for aggregates >500 μm [15]. Therefore, dynamic cultivation systems that enable enhanced mass transfer seem preferable for aggregate cultivation. Indeed, cultivation of MSC aggregates in rotating wall vessel bioreactors, shake flasks, spinner flasks, or on an orbital shaker did not result in necrotic tissue [11,16–18]. However, no study reports the cultivation in a stirred tank reactor.

Since the natural in vivo environment of MSC often displays oxygen concentrations considerably <21% O_2 (hypoxia) [19–21], the effect of reduced oxygen conditions on MSCs was extensively investigated in the past and regarding the therapeutic use of MSCs several advantages of hypoxic cultivation emerged [22]. In fact, MSCs exhibit increased proliferation [21,23,24], reduced senescence [25], and prolonged genetic stability [26], when exposed to hypoxia while maintaining their immunosuppressive properties [27].

Therefore, in the present study, we cultivated human adipose derived 3D MSCs aggregates in a continuously stirred tank reactor (CSTR) under normoxic (21% O_2) and hypoxic (5% O_2) conditions. Since rather high shear forces can occur in a stirred tank reactor, the actual shear stress was estimated via computational fluid dynamics. After cultivation, surface marker expression and differentiation capacity was evaluated.

Although average shear forces of 0.2 Pa were present, MSCs did not differentiate spontaneously and maintained their innate phenotype and trilineage differentiation potential. Cultivation of aggregates in a CSTR might be beneficial for large scale expansion of MSCs, production of MSC aggregates, or their secreted trophic factors. To our knowledge, this is the first study that demonstrates the formation and cultivation of MSC aggregates in a CSTR.

2. Materials and Methods

2.1. Bioreactor Design

For the cultivation of human MSCs, a continuously stirred tank reactor (CSTR) was created by computer-aided design (CAD) software (Solidworks, Dassault Systèmes, Stuttgart, Germany). Based on the CAD drawings, all parts of the reactor framework were manufactured from stainless steel. As an exception, the impeller (impeller diameter $d = 25$ mm) was constructed from polyether ether ketone (PEEK) (GT Labortechnik, Arnstein, Germany). The custom-made round-bottom glass vessel (vessel diameter $D = 80$ mm, round bottom radius $r = 40$ mm) and riser pipes were hand manufactured by a local glassblower (Glaspunkt, Burghausen, Germany). The reactor bottom clearance was set equally to the impeller diameter d. In order to minimize the risk of contamination of the vessel interior, the stirring shaft was further equipped with a mechanical seal (Trelleborg Sealing Solutions, Stuttgart, Germany). For measuring the oxygen content in the culture medium during hypoxic and normoxic ambient conditions, an optical oxygen sensor spot was attached to the reactor glass wall (Presens GmbH, Regensburg, Germany).

2.2. Compuitational Fluid Dyamics

The bioreactor CAD schematics were harnessed to establish a computational fluid dynamics (CFD) model from which data on flow field and shearing were collected. The CAD model files were imported into a suitable finite element method (FEM) software (Comsol Multiphysics 5.2, Comsol Multiphysics GmbH, Göttingen, Germany) and processed for subsequent FEM computations. Briefly, the material properties of the model geometry were adjusted to comply with human MSC culture medium at 37 °C (dynamic viscosity 0.765 mPa·s, density 998 kg/m^3).

For flow velocity field computations, the Comsol rotating machinery module was used to solve the resulting set of Navier–Stokes differential equations. First, the rotating model domain was defined as such by parameterizing its rotation velocity. Here, the κ–ε turbulence model was assumed and the corresponding model constants for turbulence fluid flow were set to $C_{\varepsilon 1} = 1.44$, $C_{\varepsilon 2} = 1.92$, $C_\mu = 0.09$, $\sigma_\kappa = 1.0$, $\sigma_\varepsilon = 1.3$, Von Kármán constant $K_N = 0.41$, and wall roughness $B = 5.2$. A pressure point constraint was defined at the model surface boundary to set ambient pressure conditions. Furthermore, a symmetry boundary condition was applied to the fluid–gas interface. Wall functions were defined for all remaining boundaries. An auxiliary sweep was performed in order to stabilize solver convergence. Therefore, the dynamic viscosity of the fluid was subsequently multiplied by a numerical auxiliary factor defined as *visc_fac*. For the first computation iteration, a value of *visc_fac* = 50 was set. Following, computations for the steady-state solution were performed while stepwise lowering the auxiliary factor down to a value of *visc_fac* = 1.

2.3. Cell Culture

The use of human tissue was approved by the ethics committee of the Medical University Vienna, Austria (EK Nr. 957/2011, 30 January 2013) and the donor gave written consent. Human ASCs were isolated within 3 h after surgery as described before from a female donor [21] (48 years old). Briefly, fat tissue was minced with scissors and digested with collagenase type I (Sigma Aldrich, St. Louis, MO, USA). Subsequently, multiple centrifugation and washing steps were carried out to receive the stromal vascular fraction which was then transferred to cell culture flasks. ASCs were cultivated in standard medium composed of MEM alpha (Thermo Fisher Scientific, Waltham, MA, USA), 0.5% gentamycin (Lonza, Basel, Switzerland), 2.5% human platelet lysate (PL BioScience, Aachen, Germany) and 1 U/ml heparin (Ratiopharm, Ulm, Germany) in a humidified incubator at 37 °C, 5% CO_2 and 21% (normoxic) or 5% O_2 (hypoxic). For cryo-preservation, cells were detached by accutase treatment (GE healthcare, Little Chalfont, UK) and transferred to cryo-medium composed of 77.5% αMEM, 12.5% HPL, 10% DMSO (Sigma Aldrich), and 1 U/ml heparin as described before [9] for storage in liquid nitrogen. For bioreactor cultivation, cells were thawed, expanded until passage 2, and harvested by accutase treatment. Cells for cultivation at 5% O_2 have also been isolated and subcultivated at 5% O_2 until seeding.

2.4. Bioreactor Cultivation

After steam sterilization, the CSTR was filled with PBS at 37 °C in order to calibrate the PreSens oxygen sensor. The tank was filled and emptied through one of the ports in the lid while the lid itself was kept closed at all time. After calibration, PBS was removed with a suction pump and the tank was filled with 130 mL of a 1×10^5 cells/mL single cell suspension of MSCs (13×10^6 cells total) in standard medium with 10 instead of 2.5% human platelet lysate (PL Bioscience, Aachen, Germany) was used. Cells were cultivated for six days at 600 revolutions per minute (rpm), 37 °C, 5% CO_2, and 21% or 5% O_2. After three days, 100 mL of the medium was replaced. Aggregates and cells were allowed to sediment for 15 min prior to medium change.

After six days, the medium containing cells and aggregates was transferred to 50 mL centrifugation tubes, the tank was rinsed with 40 mL PBS, the PBS added to the tubes and the tubes centrifuged for 5 min at 500× *g*. The bioreactor tank was filled with 25 mL of a 37 °C pre-warmed Accumax solution (Sigma Aldrich, St. Louis, MS, USA) and incubated 15 min at 37 °C in order to remove adherent cells from the glass wall. In parallel, supernatant from centrifugation tubes was removed, the pellets resuspended in 40 mL PBS, unified in one tube, and again centrifuged for 5 min at 500× *g*. After removal of the supernatant, the pellet was resuspended in Accumax solution from the bioreactor tank. After this, the solution was incubated 15 min in a 37 °C water bath and further for 30 min on a horizontal shaker at 300 rpm and 37 °C in order to dissociate the aggregates. Then, the cell suspension was passed through a cell strainer to separate single cells from remaining aggregates. Cells were counted by trypan blue staining with a hemocytometer after incubation to determine cell number and viability (overall cells were incubated for 45 min in Accumax solution). Furthermore, the cell strainer was placed in a 6-well plate and incubated in 6 mL Accumax solution for 1 h at 100 rpm and 37 °C. Single cells released from the aggregates were also counted by trypan blue staining. Harvested ASCs were frozen as described above for further analysis of surface markers and differentiation capacity.

2.5. Phenotyping

To determine MSC surface marker expression cells were detached by accutase treatment and stained with MSC phenotyping kit (Miltenyi Biotech GmbH, Bergisch Gladbach, Germany) according to manufacturer's instructions. Stained cells (5×10^5 cells per aliquot) were resuspended in 300 µL flow cytometry buffer and acquisition was carried out on a Gallios flow cytometer (Beckman Coulter, Brea, CA, USA). Between $1–5 \times 10^4$ gated events were recorded. Subsequent analysis was performed with Kaluza Flow Cytometry software (version 1.3, Beckman Coulter, Brea, CA, USA).

2.6. Differentiation

To evaluate the differentiation capacity of ASCs after cultivation in the CSTR, cells were thawed and cultivated in cell culture flasks to approximately 80% confluency. Subsequently, cells were detached by accutase treatment and seeded into fibronectin coated 12-well plates (BD Bioscience, Franklin Lakes, NJ, USA) at a density of 4000 c/cm^2. When cells reached confluency, the medium was changed to adipogenic, chondrogenic (both Miltenyi Biotec GmbH, Bergisch Gladbach, Germany), or osteogenic medium (standard medium supplemented with 5 mM beta-glycerolphosphate, 0.1 µM Dexamethasone, 200 µM L-ascorbate-2-phosphate, all from Sigma Aldrich, St. Louis, USA) respectively. Cells were cultivated for 21 days and medium was changed every 2–3 days. Afterwards, cells from chondrogenic and osteogenic differentiation were fixated with 96% ethanol while cells from adipogenic differentiation were fixated with 4% paraformaldehyde for further histological staining.

2.7. Histologic Stainings

ASCs cultivated in adipogenic medium were stained with Oil Red O (staining of lipid vacuoles). For this, cells were rinsed with ddH$_2$O and incubated in Oil Red O solution (Sigma Aldrich) for 20 min. ASCs cultivated in chondrogenic medium were stained with alcian blue (staining of glycosaminoglycans). Briefly, cells were rinsed with 3% acetic acid and incubated in alcian blue solution (1% *w/v* alcian blue in 3% acetic acid) for 30 min. ASCs cultivated in osteogenic medium were double stained with DAPI (Sigma Aldrich) and calcein (staining for calcium, Franklin) and silver nitrate (also known as von Kossa; staining for phosphates). For fluorescent double staining, cells were rinsed with PBS and incubated in DAPI solution (4 µL/mL DAPI in PBS). Subsequently, cells were rinsed with ddH$_2$O and incubated in calcein solution (5 µg/mL) over night at 4 °C. Cells for staining of phosphates were rinsed with ddH$_2$O and incubated in 5% silver nitrate solution (Carl Roth) for 30 min in the dark. Subsequently, cells were rinsed again, exposed to UV light for 2 min, and rinsed with decolorization solution (5% Na$_2$CO$_3$, 0.2% formaldehyde in ddH$_2$O).

2.8. Statistical Analysis

All results are presented as mean ± standard deviation (SD). Comparisons were carried out by the unpaired, two-sided *t*-test. Values of $p < 0.1$ with a confidence interval of 90% were defined as statistically significant (as indicated by an asterisk). All analysis were carried out with GraphPad Prism 6.01 (GraphPad Software, Inc., La Jolla, CA, USA).

3. Results

3.1. Shear Stress Estimation By Computational Fluid Dynamics

The impeller of the continuously stirred bioreactor was designed to provide gentle, yet thorough mixing at low shear when operated at moderate impeller speed in the range of 100–120 rpm. However, in order to obviate cells from attaching to the bioreactor glass wall, vigorous stirring rates were necessary. The increase of the impeller rotation frequency to 600 rpm leads to considerable magnitudes of shear stress. To evaluate the impact of high rotational velocities, computational fluid dynamics were used to compute the fluid flow regime of the continuously stirred bioreactor. Here, as shown by the color legend in Figure 1, the calculated shear stress is predominantly found in the range between 0.05 Pa and 0.35 Pa. A closer investigation of the computational fluid dynamics results revealed peak stress levels of 2.5 Pa at the impeller blade tips. When considering the entire spatial dimension of the bioreactor, 0.02 Pa were obtained as average shear stress in total.

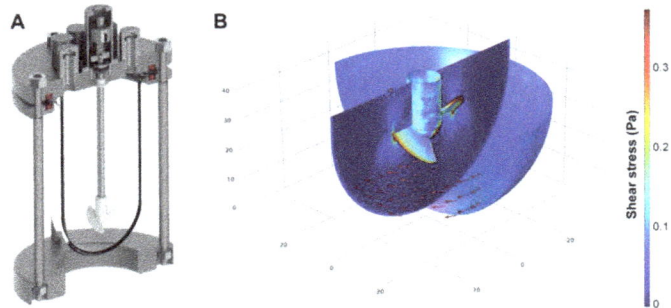

Figure 1. (**A**) Three-dimensional model of the stirred tank reactor used for aggregate cultivation; (**B**) Flow field direction (red arrows) and shear stress distribution (color legend) at a rotational speed of 600 rpm as estimated by computational fluid dynamics.

3.2. Bioreactor Cultivation

Human ASCs were cultivated for 6 days at 21% or 5% O_2 in a continuously stirred tank reactor. Visible aggregates formed spontaneously after approximately three days. Cells and aggregates were also found to adhere and grow on the glass surface of the bioreactor vessel and partially on the impeller. Dissolved oxygen (DO) decreased slowly under normoxic conditions to approximately 85% until day 6 whereas under hypoxic conditions it decreased to 0% after five days (Figure 2). Cells expanded 1.85-fold ($\pm$0.19) under normoxic conditions and 2.23-fold ($\pm$0.27) under hypoxic conditions displaying a viability of 78.5 $\pm$ 9.8% and 86 $\pm$ 3.1% respectively. Human MSCs of different origin displayed an approximately 1.3-fold increased growth rate when cultivated under hypoxic conditions [28]. Although not statistically significant, data of the present study indicate a similar behavior when ASCs are cultivated in a CSTR. Furthermore, glucose consumption (0.85 $\pm$ 0.1 mmol) and lactate production (1.69 $\pm$ 0.11 mmol) were significantly lower in normoxic conditions compared to hypoxic conditions, where glucose consumption was 1.09 $\pm$ 0.02 mmol and lactate production 2.05 $\pm$ 0.09 mmol. In the absence of oxygen, glycolytic activity increases since glucose is metabolized rather by lactate acid fermentation than by oxidative phosphorylation in the mitochondria, which reduces the efficiency of ATP production. However, under hypoxic cultivation, cell numbers were slightly increased together with a higher viability (Figure 3).

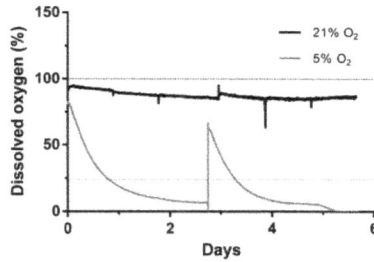

Figure 2. Dissolved oxygen during cultivation mesenchymal stem cells in continuously stirred tank reactor at 21% or 5% O_2 ambient oxygen. Dashed lines indicate the respective dissolved oxygen concentration at respective equilibrium.

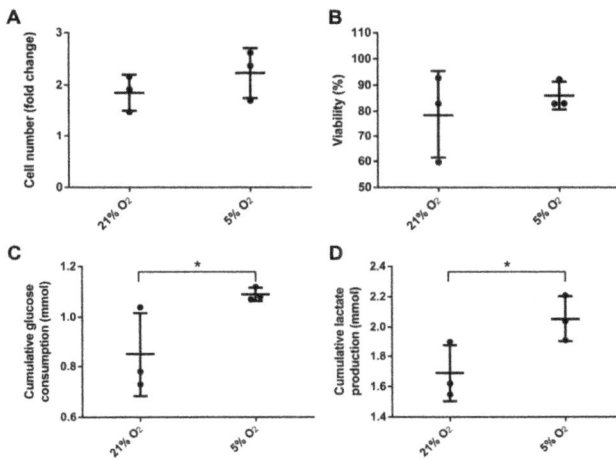

Figure 3. (**A**) Yield of viable cells; (**B**) overall viability, cumulative (**C**) glucose consumption and (**D**) lactate production of mesenchymal stem cells after six days cultivation under 21% and 5% O_2 in a continuously stirred tank reactor ($n = 3$). Data is represented as mean $\pm$ SD, asterisks indicate statistically significant difference ($p < 0.1$, confidence interval of 90%).

3.3. Stem Cell Properties

Maintaining stem cell properties during ex vivo cultivation is mandatory in the context of stem cell expansion for later use in cell-based therapies. Stem cell properties were evaluated by antibody staining of characteristic surface markers that meet the minimal criteria of MSC and evaluation of the differentiation capacity [2]. Surface markers of MSCs before and after cultivation in the CSTR were comparable (Figure 4). Also, surface markers of cells cultivated under normoxic and hypoxic conditions were comparable and met the minimal criteria of MSCs. Furthermore, differentiation into adipogenic, chondrogenic, and osteogenic lineage was observed (Figure 5). However, slightly elevated adipogenic and chondrogenic but a reduced osteogenic differentiation was observed in hypoxic conditions compared to normoxic in the present study.

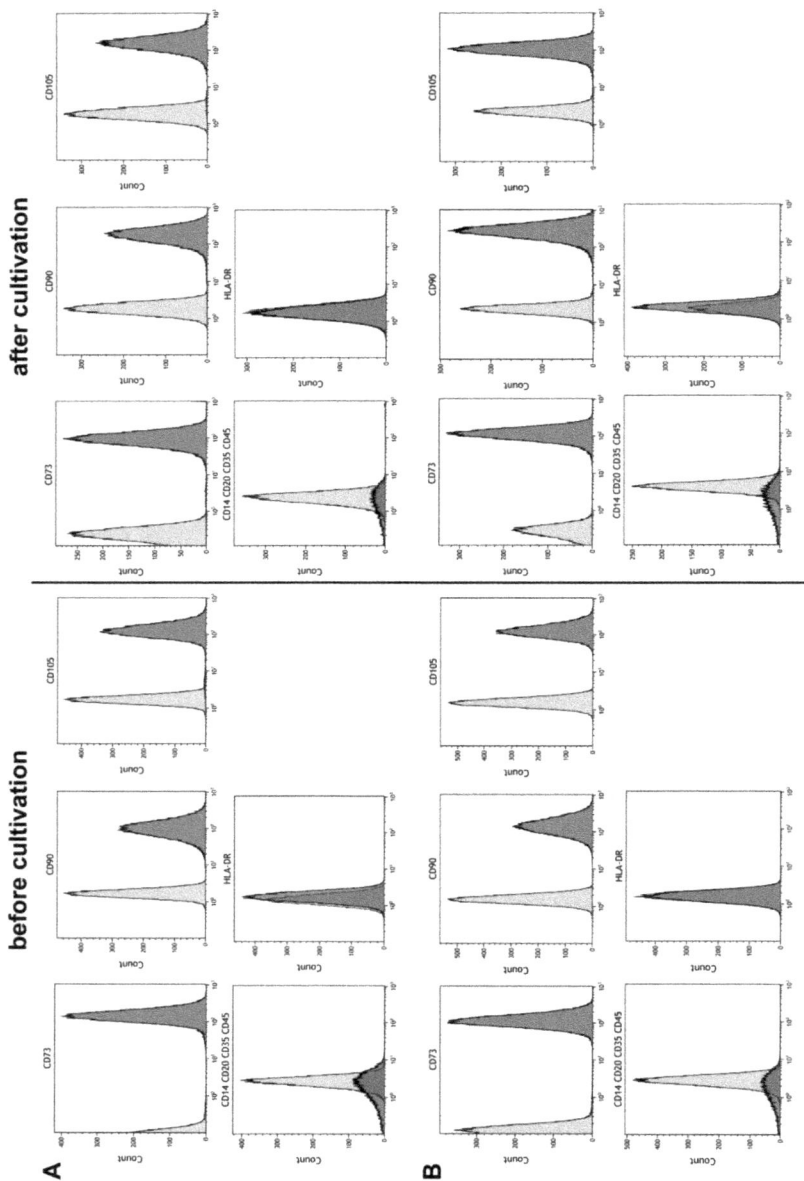

Figure 4. Phenotyping of mesenchymal stem cells before and after six days of cultivation under (**A**) 21% and (**B**) 5% O_2 in a continuously stirred tank reactor. Light gray areas indicate the isotype control, dark grey areas indicate the phenotype.

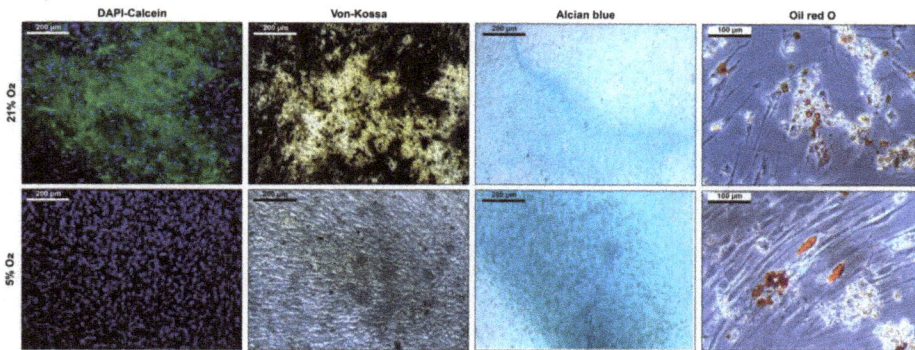

Figure 5. Differentiation of mesenchymal stem cells after six days cultivation under 21% and 5% O_2 in a continuously stirred tank reactor. Osteogenic differentiation is indicated by DAPI-calcein staining for nuclei and extracellular calcium and von Kossa stain for extracellular phosphates. Chondrogenic differentiation is indicated by alcian blue staining which stains for glycosaminoglycans. Adipogenic differentiation is indicated by Oil Red O staining which stains the intracellular fatty vacuoles.

4. Discussion

Cultivation of MSCs aggregates or spheroids is well-established at the scale of hanging drops [14], microtiter plates [29], or shaking flasks [17,30]. The maintenance of stem cell properties after cultivation on an orbital shaker was already demonstrated in a study with murine MSC aggregates [18]. Another study reported the cultivation of human MSCs in shaking flasks [17]. However, to the best of our knowledge, no reports have been made regarding the upscale of MSC aggregate cultivation in a stirred tank reactor. Therefore, in this study, we investigated the cultivation of MSC aggregates in a CSTR reactor under normoxic (21% O_2) and hypoxic (5% O_2) conditions. Cells cultivated under hypoxic conditions displayed increased proliferation, viability (not significant), and glycolytic activity (significant) compared to normoxic conditions. Furthermore, MSCs maintained their stem cell properties as indicated by their immunophenotype (positive for CD73, CD90, CD105 and negative for CD14, CD20, CD35, CD45, and HLA-DR) and multilineage differentiation capacity. To avoid cell adhesion on the glass wall and impeller of the bioreactor the impeller speed was set to 600 rpm. The resulting average shear stress, as estimated by computational fluid dynamics, was found to be 0.02 Pa with peak shear stress of 2.5 Pa. MSCs are known to react to mechanical cues such as fluid shear forces and several studies reported differentiation, when cells were exposed to shear stress as low as 7.6×10^{-5} [31] or 0.01 Pa [32]. However, flow cytometry analysis of surface marker expression revealed exclusively undifferentiated stem cells as no side populations were found. Cellular aggregates were shown to be temporally and spatially heterogeneous with regards to cellular architecture and expression of matrix proteins [11,30,33]. Therefore, cells on the aggregate surface might shield the inner cell mass from fluid shearing. Furthermore, cells maintained their trilineage differentiation potential and displayed slightly elevated adipogenic and chondrogenic but a reduced osteogenic differentiation. The majority of studies reported attenuated osteogenic [34] and adipogenic [28], but elevated chondrogenic [26], differentiation under hypoxic conditions although elevated adipogenic differentiation was also observed [35].

The traditional approach for the expansion of MSCs is the cultivation on two-dimensional plastic surfaces. Although this approach is simple and reproducible it is time and material consuming to achieve high cell numbers. Also, a rather high surface is required which often results in numerous cell culture vessels and extensive passaging of cells. Due to the comparably high number of opening events the risk of contamination is increased. Therefore, other approaches like the cultivation in hollow-fiber bioreactors with a high surface-to-volume ratio or microcarrier-based systems were developed. For example the quantum system by Terumo expanded 6.6×10^8 MSCs from a 25 mL bone marrow aspirate [36]. Also, expansion of MSCs on microcarrier was performed successfully in spinner flasks [37] and CSTRs [38]. However, the quantum system was mainly developed for use at a clinical scale, and thus might be oversized for research scale while cell harvest is challenging in microcarrier-based systems [39,40]. Also, both approaches are limited by their surface capacity. Though the yield of the presented scaffold-free aggregate cultivation of MSCs is comparably low this system offers a straight forward approach to cultivation of MSCs. Potentially, advantages of traditional suspension culture, such as simplicity, standardizability, rapid development of cultivation protocols, and upscale are translatable to aggregate cultivation. Nevertheless, since homogeneity of aggregates is not given until now, the system may be of interest rather for the production of secreted compounds than cell expansion. Especially, aggregate cultivation under hypoxic conditions might be of benefit in this context since the hypoxic environment in the core of aggregates was reported to increase expression of trophic factors such as VEGF, FGF-2, HGF, and CXCR4 [16,41] as well as ECM proteins such as fibronectin, collagen I, vitronectin, and collagen IV [42]. However, the expression profile of MSC aggregates cultivated in an extensively stirred environment needs further investigation.

In order to improve the cultivation process with regards to yield, viability, and reproducibility further work will focus on the optimization of the bioreactor and impeller geometry. Furthermore, coating agents like silicon will be considered to prevent cell loss due to glass adherence. This might also reduce the required impeller speed and thus contribute to proliferation and viability of cells. Cultivation under hypoxic conditions seems to improve cell yield and viability. However, an oxygen concentration of $\leq 12\%$ DO ($\approx 2.5\%$ ambient O_2) should be avoided as it was shown to decrease proliferation of MSCs [21]. Also, serum-free medium was optimized for MSC aggregate cultivation and shown to increase proliferation and similar or enhanced differentiation capacity compared to cultivation in serum-containing medium [18].

5. Conclusions

We demonstrated the cultivation of 3D MSC aggregates in a CSTR under normoxic and hypoxic conditions. The cultivation under hypoxic conditions resulted in slightly higher yield and viability of cells. Although the exerted shear forces were comparatively high, MSCs maintained their immunophenotype and differentiation capacity. We hypotesize cultivation of MSC aggregates in a CSTR is a viable option for expansion of MSCs or production of secreted compounds after further optimization of cultivation conditions.

Author Contributions: D.E. designed and performed cell culture experiments and data analysis, and prepared the manuscript; I.S. designed the continuously stirred tank reactor (CSTR), performed computational analysis, and prepared the manuscript; J.H. prepared the manuscript; C.K. designed the study and prepared the manuscript.

Conflicts of Interest: The authors declare no conflict of interest.

References

1. Pittenger, M.F.; Mackay, A.M.; Beck, S.C.; Jaiswal, R.K.; Douglas, R.; Mosca, J.D.; Moorman, M.A.; Simonetti, D.W.; Craig, S.; Marshak, D.R. Multilineage Potential of Adult Human Mesenchymal Stem Cells. *Science* **1999**, *284*, 143–147. [CrossRef] [PubMed]
2. Dominici, M.; Le Blanc, K.; Mueller, I.; Slaper-Cortenbach, I.; Marini, F.; Krause, D.; Deans, R.; Keating, A.; Prockop, D.; Horwitz, E. Minimal criteria for defining multipotent mesenchymal stromal cells. The International Society for Cellular Therapy position statement. *Cytotherapy* **2006**, *8*, 315–317. [CrossRef] [PubMed]
3. Oh, J.Y.; Kim, M.K.; Shin, M.S.; Lee, H.J.; Ko, J.H.; Wee, W.R.; Lee, J.H. The Anti-Inflammatory and Anti-Angiogenic Role of Mesenchymal Stem Cells in Corneal Wound Healing Following Chemical Injury. *Stem Cells* **2008**, *26*, 1047–1055. [CrossRef] [PubMed]

4. Capla, A.I. Adult Mesenchymal Stem Cells for Tissue Engineering Versus Regenerative Medicine. *J. Cell. Physiol.* **2007**, 341–347. [CrossRef] [PubMed]
5. Teixeira, F.G.; Carvalho, M.M.; Sousa, N.; Salgado, A.J. Mesenchymal stem cells secretome: A new paradigm for central nervous system regeneration? *Cell. Mol. Life Sci.* **2013**, *70*, 3871–3882. [CrossRef] [PubMed]
6. Ranganath, S.H.; Levy, O.; Inamdar, M.S.; Karp, J.M. Harnessing the mesenchymal stem cell secretome for the treatment of cardiovascular disease. *Cell. Stem Cell.* **2012**, *10*, 244–258. [CrossRef] [PubMed]
7. Galipeau, J. The mesenchymal stromal cells dilemma-does a negative phase III trial of random donor mesenchymal stromal cells in steroid-resistant graft-versus-host disease represent a death knell or a bump in the road? *Cytotherapy* **2013**, *15*, 2–8. [CrossRef] [PubMed]
8. Von Bahr, L.; Sundberg, B.; Lonnies, L.; Sander, B.; Karbach, H.; Hagglund, H.; Ljungman, P.; Gustafsson, B.; Karlsson, H.; Le Blanc, K.; et al. Long-term complications, immunologic effects, and role of passage for outcome in mesenchymal stromal cell therapy. *Biol. Blood Marrow Transplant.* **2012**, *18*, 557–564. [CrossRef] [PubMed]
9. Mitchell, J.B.; McIntosh, K.; Zvonic, S.; Garrett, S.; Floyd, Z.E.; Kloster, A.; Di Halvorsen, Y.; Storms, R.W.; Goh, B.; Kilroy, G.; et al. Immunophenotype of Human Adipose-Derived Cells: Temporal Changes in Stromal-Associated and Stem Cell-Associated Markers. *Stem Cells* **2006**, *24*, 376–385. [CrossRef] [PubMed]
10. Bartosh, T.J.; Ylostalo, J.H.; Mohammadipoor, A.; Bazhanov, N.; Coble, K.; Claypool, K.; Lee, R.H.; Choi, H.; Prockop, D.J. Aggregation of human mesenchymal stromal cells (MSCs) into 3D spheroids enhances their antiinflammatory properties. *Proc. Natl. Acad. Sci.* **2010**, *107*, 13724–13729. [CrossRef] [PubMed]
11. Frith, J.E.; Thomson, B.; Genever, P.G. Dynamic Three-Dimensional Culture Methods Enhance Mesenchymal Stem Cell Properties and Increase Therapeutic Potential. *Tissue Eng. Part. C Methods* **2010**, *16*, 735–749. [CrossRef] [PubMed]
12. Ylöstalo, J.H.; Bartosh, T.J.; Coble, K.; Prockop, D.J. Human mesenchymal stem/stromal cells cultured as spheroids are self-activated to produce prostaglandin E2 that directs stimulated macrophages into an anti-inflammatory phenotype. *Stem Cells* **2012**, *30*, 2283–2296. [CrossRef] [PubMed]
13. Cheng, N.-C.; Chen, S.-Y.; Li, J.-R.; Young, T.-H. Short-Term Spheroid Formation Enhances the Regenerative Capacity of Adipose-Derived Stem Cells by Promoting Stemness, Angiogenesis, and Chemotaxis. *Stem Cells Transl. Med.* **2013**, *2*, 584–594. [CrossRef] [PubMed]
14. Hildebrandt, C.; Büth, H.; Thielecke, H. A scaffold-free in vitro model for osteogenesis of human mesenchymal stem cells. *Tissue Cell* **2011**, *43*, 91–100. [CrossRef] [PubMed]
15. Alvarez-Pérez, J.; Ballesteros, P.; Cerdán, S. Microscopic images of intraspheroidal pH by 1H magnetic resonance chemical shift imaging of pH sensitive indicators. *Magn. Reson. Mater. Physics, Biol. Med.* **2005**, *18*, 293–301. [CrossRef] [PubMed]
16. Bhang, S.H.; Cho, S.W.; La, W.G.; Lee, T.J.; Yang, H.S.; Sun, A.Y.; Baek, S.H.; Rhie, J.W.; Kim, B.S. Angiogenesis in ischemic tissue produced by spheroid grafting of human adipose-derived stromal cells. *Biomaterials* **2011**, *32*, 2734–2747. [CrossRef] [PubMed]
17. Alimperti, S.; Lei, P.; Wen, Y.; Tian, J.; Campbell, A.M.; Andreadis, S.T. Serum-free spheroid suspension culture maintains mesenchymal stem cell proliferation and differentiation potential. *Biotechnol. Prog.* **2014**, *30*, 974–983. [CrossRef] [PubMed]
18. Baraniak, P.R.; McDevitt, T.C. Scaffold-free culture of mesenchymal stem cell spheroids in suspension preserves multilineage potential. *Cell. Tissue Res.* **2012**, *347*, 701–711. [CrossRef] [PubMed]
19. Harrison, J.S.; Rameshwar, P.; Chang, V.; Bandar, P. Oxygen saturation in the bone marrow of healthy volunteers. *Blood* **2002**, *125*, 1679–1682. [CrossRef]
20. Chow, D.C.; Wenning, L.A.; Miller, W.M.; Papoutsakis, E.T. Modeling pO_2 Distributions in the Bone Marrow Hematopoietic Compartment. II. Modified Kroghian Models. *Biophys. J.* **2001**, *81*, 685–696. [CrossRef]
21. Lavrentieva, A.; Majore, I.; Kasper, C.; Hass, R. Effects of hypoxic culture conditions on umbilical cord-derived human mesenchymal stem cells. *Cell. Commun. Signal.* **2010**, *8*, 18. [CrossRef] [PubMed]
22. Buravkova, L.B.; Andreeva, E.R.; Gogvadze, V.; Zhivotovsky, B. Mesenchymal stem cells and hypoxia: Where are we? *Mitochondrion* **2014**, *19*, 105–112. [CrossRef] [PubMed]
23. Iida, K.; Takeda-Kawaguchi, T.; Tezuka, Y.; Kunisada, T.; Shibata, T.; Tezuka, K.I. Hypoxia enhances colony formation and proliferation but inhibits differentiation of human dental pulp cells. *Arch. Oral Biol.* **2010**, *55*, 648–654. [CrossRef] [PubMed]

24. Grayson, W.L.; Zhao, F.; Bunnell, B.; Ma, T. Hypoxia enhances proliferation and tissue formation of human mesenchymal stem cells. *Biochem. Biophys. Res. Commun.* **2007**, *358*, 948–953. [CrossRef] [PubMed]
25. Tsai, C.C.; Chen, Y.J.; Yew, T.L.; Chen, L.L.; Wang, J.Y.; Chiu, C.H.; Hung, S.C. Hypoxia inhibits senescence and maintains mesenchymal stem cell properties through down-regulation of E2A-p21 by HIF-TWIST. *Blood* **2011**, *117*, 459–469. [CrossRef] [PubMed]
26. Estrada, J.C.; Albo, C.; Benguría, A.; Dopazo, A.; López-Romero, P.; Carrera-Quintanar, L.; Roche, E.; Clemente, E.P.; Enríquez, J.A.; Bernad, A.; et al. Culture of human mesenchymal stem cells at low oxygen tension improves growth and genetic stability by activating glycolysis. *Cell. Death Differ.* **2012**, *19*, 743–755. [CrossRef] [PubMed]
27. Nold, P.; Hackstein, H.; Riedlinger, T.; Kasper, C.; Neumann, A.; Mernberger, M.; Fölsch, C.; Schmitt, J.; Fuchs-Winkelmann, S.; Barckhausen, C.; et al. Immunosuppressive capabilities of mesenchymal stromal cells are maintained under hypoxic growth conditions and after gamma irradiation. *Cytotherapy* **2015**, *17*, 152–162. [CrossRef] [PubMed]
28. Dos Santos, F.; Andrade, P.Z.; Boura, J.S.; Abecasis, M.M.; Da Silva, C.L.; Cabral, J.M.S. Ex vivo expansion of human mesenchymal stem cells: A more effective cell proliferation kinetics and metabolism under hypoxia. *J. Cell. Physiol.* **2010**, *223*, 27–35. [CrossRef] [PubMed]
29. Langenbach, F.; Berr, K.; Naujoks, C.; Hassel, A.; Hentschel, M.; Depprich, R.; Kubler, N.R.; Meyer, U.; Wiesmann, H.-P.; Kögler, G.; et al. Generation and differentiation of microtissues from multipotent precursor cells for use in tissue engineering. *Nat. Protoc.* **2011**, *6*, 1726–1735. [CrossRef] [PubMed]
30. Sart, S.; Tsai, A.-C.; Li, Y.; Ma, T. Three-Dimensional Aggregates of Mesenchymal Stem Cells: Cellular Mechanisms, Biological Properties, and Applications. *Tissue Eng. Part. B Rev.* **2014**, *20*, 365–380. [CrossRef] [PubMed]
31. Kleinhans, C.; Mohan, R.R.; Vacun, G.; Schwarz, T.; Haller, B.; Sun, Y.; Kahlig, A.; Kluger, P.; Finne-Wistrand, A.; Walles, H.; et al. A perfusion bioreactor system efficiently generates cell-loaded bone substitute materials for addressing critical size bone defects. *Biotechnol. J.* **2015**, *10*, 1727–1738. [CrossRef] [PubMed]
32. Egger, D.; Spitz, S.; Fischer, M.; Handschuh, S.; Glösmann, M.; Friemert, B.; Egerbacher, M.; Kasper, C. Application of a Parallelizable Perfusion Bioreactor for Physiologic 3D Cell Culture. *Cells Tissues Organs* **2017**, *203*, 316–326. [CrossRef] [PubMed]
33. Baraniak, P.R.; Cooke, M.T.; Saeed, R.; Kinney, M.A.; Fridley, K.M.; McDevitt, T.C. Stiffening of human mesenchymal stem cell spheroid microenvironments induced by incorporation of gelatin microparticles. *J. Mech. Behav. Biomed. Mater.* **2012**, *11*, 63–71. [CrossRef] [PubMed]
34. Malladi, P.; Xu, Y.; Chiou, M.; Giaccia, A.J.; Michael, T.; Longaker, M.T. Effect of reduced oxygen tension on chondrogenesis and osteogenesis in adipose-derived mesenchymal cells. *AJP Cell Physiol.* **2006**, *290*, C1139–C1146. [CrossRef] [PubMed]
35. Valorani, M.G.; Montelatici, E.; Germani, A.; Biddle, A.; D'Alessandro, D.; Strollo, R.; Patrizi, M.P.; Lazzari, L.; Nye, E.; Otto, W.R.; et al. Pre-culturing human adipose tissue mesenchymal stem cells under hypoxia increases their adipogenic and osteogenic differentiation potentials. *Cell. Prolif.* **2012**, *45*, 225–238. [CrossRef] [PubMed]
36. Hanley, P.J.; Mei, Z.; Durett, A.G.; da Graca Cabreira-Harrison, M.; Klis, M.; Li, W.; Zhao, Y.; Yang, B.; Parsha, K.; Mir, O.; et al. Efficient manufacturing of therapeutic mesenchymal stromal cells with the use of the Quantum Cell Expansion System. *Cytotherapy* **2014**, *16*, 1048–1058. [CrossRef] [PubMed]
37. Eibes, G.; dos Santos, F.; Andrade, P.Z.; Boura, J.S.; Abecasis, M.M.A.; da Silva, C.L.; Cabral, J.M.S. Maximizing the ex vivo expansion of human mesenchymal stem cells using a microcarrier-based stirred culture system. *J. Biotechnol.* **2010**, *146*, 194–197. [CrossRef] [PubMed]
38. Dos Santos, F.; Campbell, A.; Fernandes-Platzgummer, A.; Andrade, P.Z.; Gimble, J.M.; Wen, Y.; Boucher, S.; Vemuri, M.C.; Da Silva, C.L.; Cabral, J.M.S. A xenogeneic-free bioreactor system for the clinical-scale expansion of human mesenchymal stem/stromal cells. *Biotechnol. Bioeng.* **2014**, *111*, 1116–1127. [CrossRef] [PubMed]
39. Liu, N.; Zang, R.; Yang, S.T.; Li, Y. Stem cell engineering in bioreactors for large-scale bioprocessing. *Eng. Life Sci.* **2014**, *14*, 4–15. [CrossRef]
40. Nienow, A.W.; Rafiq, Q.A.; Coopman, K.; Hewitt, C.J. A potentially scalable method for the harvesting of hMSCs from microcarriers. *Biochem. Eng. J.* **2014**, *85*, 79–88. [CrossRef]

Bioengineering **2017**, *4*, 47

41. Zhang, Q.; Nguyen, A.L.; Shi, S.; Hill, C.; Wilder-Smith, P.; Krasieva, T.B.; Le, A.D. Three-Dimensional Spheroid Culture of Human Gingiva-Derived Mesenchymal Stem Cells Enhances Mitigation of Chemotherapy-Induced Oral Mucositis. *Stem Cells Dev.* **2012**, *21*, 937–947. [CrossRef] [PubMed]
42. Kim, J.; Ma, T. Autocrine fibroblast growth factor 2-mediated interactions between human mesenchymal stem cells and the extracellular matrix under varying oxygen tension. *J. Cell. Biochem* **2013**, *114*, 716–727. [CrossRef] [PubMed]

bioengineering

MDPI

Article

Development and Characterization of a Parallelizable Perfusion Bioreactor for 3D Cell Culture

Dominik Egger [1], Monica Fischer [1], Andreas Clementi [1], Volker Ribitsch [2], Jan Hansmann [3] and Cornelia Kasper [1,*]

[1] Department of Biotechnology, University of Natural Resources and Life Sciences, Muthgasse 18, 1190 Vienna, Austria; dominik.egger@boku.ac.at (D.E.); monica.fischer@gmx.at (M.F.); andreas.clementi@boku.ac.at (A.C.)

[2] Institute of Chemistry, University of Graz, Heinrichstraße 28/IV, 8010 Graz, Austria; volker.ribitsch@uni-graz.at

[3] Translational Center, University Hospital Wurzburg, Roentgenring 11, 97070 Wuerzburg, Germany; jan.hansmann@uni-wuerzburg.de

* Correspondence: cornelia.kasper@boku.ac.at; Tel.: +43-1-47654-79030

Academic Editor: Christoph Herwig
Received: 29 March 2017; Accepted: 23 May 2017; Published: 25 May 2017

Abstract: The three dimensional (3D) cultivation of stem cells in dynamic bioreactor systems is essential in the context of regenerative medicine. Still, there is a lack of bioreactor systems that allow the cultivation of multiple independent samples under different conditions while ensuring comprehensive control over the mechanical environment. Therefore, we developed a miniaturized, parallelizable perfusion bioreactor system with two different bioreactor chambers. Pressure sensors were also implemented to determine the permeability of biomaterials which allows us to approximate the shear stress conditions. To characterize the flow velocity and shear stress profile of a porous scaffold in both bioreactor chambers, a computational fluid dynamics analysis was performed. Furthermore, the mixing behavior was characterized by acquisition of the residence time distributions. Finally, the effects of the different flow and shear stress profiles of the bioreactor chambers on osteogenic differentiation of human mesenchymal stem cells were evaluated in a proof of concept study. In conclusion, the data from computational fluid dynamics and shear stress calculations were found to be predictable for relative comparison of the bioreactor geometries, but not for final determination of the optimal flow rate. However, we suggest that the system is beneficial for parallel dynamic cultivation of multiple samples for 3D cell culture processes.

Keywords: perfusion bioreactor system; 3D cell culture; dynamic cultivation; fluid shear stress; computational fluid dynamics

1. Introduction

During the last decade, the importance of three dimensional (3D) cultivation of stem cells in dynamic bioreactor systems for tissue engineering processes, biomaterial testing, and in vitro models became very important. Conventional two dimensional (2D) static cultivation systems are used in many studies, although they do not represent the in vivo situation. Moreover, static systems have disadvantages in the mass transport of nutrients and oxygen into 3D constructs [1]. To overcome these drawbacks, different bioreactors have been developed ranging from spinner flasks [2–4], stirred systems [5,6], rotating wall [7] and rotating bed [8,9], to perfusion bioreactors [10–13] and also microfluidic systems [14,15].

While spinner flasks, stirred systems, and rotating wall reactors provide for a homogeneous distribution of nutrients in the bioreactor chamber, the mass transport into a 3D cell-scaffold construct is limited. In contrast, perfusion bioreactors force the fluid to actively pass through the scaffold in order to enhance mass transport and avoid concentration gradients [13]. Moreover, they can be used to control the mechanical environment via the application of fluid shear forces or hydrostatic pressure [16,17]. Several perfusion systems have been developed and proved to be beneficial for various 3D cell culture purposes [18–21]. However, currently available systems often have limitations and drawbacks. Most of the systems lack the possibility to cultivate multiple independent samples at once under different conditions. Therefore, the optimization of cell culture conditions is time consuming, costly, and the results may not be reproducible. Online monitoring and recording of cell culture parameters (O_2, CO_2, and temperature) is often neglected and valuable information is not available. In addition, most of the systems are not flexible with regards to scaffold size and stiffness, bioreactor chamber and tubing material, or programming of different flow regimes (e.g., alternating or intermittent flow regimes).

Mechanical stimuli such as compression, tension, hydrostatic pressure, or fluid shear forces influence stem cell behavior and often support differentiation towards a specific lineage [16,17,22]. Consequently, comprehensive knowledge and control over the fluid shear forces in a perfusion bioreactor is desirable. When the bioreactor system and material is well characterized and thus inherent parameters are known, integrated pressure sensors can be used to estimate and control fluid shear forces during cultivation.

Thus, in this study we developed a miniaturized perfusion bioreactor system that together with a tailor-made incubator system is flexible, modular, and parallelizable (up to 16 bioreactors). While a previous study focused on the application of this system [23], this study concentrates on the technical characterization. Implemented pressure sensors were characterized and used to determine the permeability of a porous scaffold which is an important material characteristic to predict shear stress. Furthermore, two bioreactor chambers with different geometries were designed and manufactured. Computational simulations were conducted together with pressure measurements and wash out experiments to characterize and compare the bioreactors for their flow velocity profile, residence time distribution, and shear stress conditions. In a proof of concept study, both bioreactor chambers were compared and the entire system was investigated for its suitability for a bone tissue engineering process. For this, human adipose derived mesenchymal stem cells (ASCs) were seeded on the previously characterized porous scaffold and cultivated for 21 days. The flow rate was set to generate shear stress conditions in a physiologic range according to prior shear stress estimations by mathematical models and CFD simulations of both bioreactor chambers.

2. Materials and Methods

2.1. Bioreactor Design

For the development of a perfusion bioreactor chamber, two different prototypes were designed with the help of computer aided design (CAD). Both the chambers were shaped to be suitable for biomaterials that were 10 mm in diameter (Figure 1). The first bioreactor (BR1) was constructed as simply as possible consisting of only a piston, a housing, and two Luer lock screws made from stainless steel, as described before [24]. The inner flow channel of the piston was set to a diameter of 3 mm and the maximum scaffold thickness was limited to 12 mm. A fluorelastomer tubing system (VWR, Darmstadt, Germany) with an inner diameter of 1.6 mm connected the bioreactor chamber to a medium reservoir via Luer lock connectors (Pieper Filter, Bad Zwischenahn, Germany). Sealing rings made of ethylene propylene diene monomer (EPDM) 70 (COG, Pinneberg, Germany) were used for sealing of the inner part and the Luer lock connections.

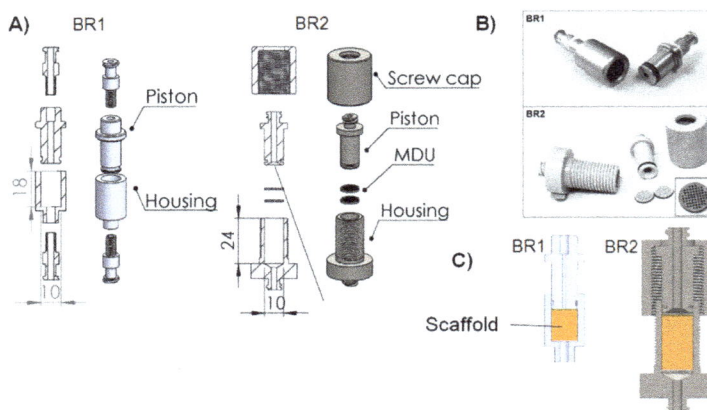

Figure 1. (**A**) Explosion view, (**B**) cross section with inserted biomaterial and (**C**) pictures of the bioreactor chamber prototypes. The first prototype (BR1) was made of stainless steel whereas the second prototype (BR2) made of polyoxymethylene (POM) features two sieve-like medium distribution units (MDU). The flow channel of BR2 opens up to the full diameter of the chamber (detailed view) and the chamber is adjustable in height to allow scaffolds of different thickness to be inserted. The placement of the scaffold in both bioreactors is represented with a yellow box in panel B.

With the second prototype (BR2), several improvements were implemented. Sieve-like medium distribution units (MDU) hold the matrices in place and support proper medium distribution. To force the liquid to flow through the MDUs, they are 10 mm in diameter and sealed with an EPDM 70 ring. Furthermore, 46 pores with a diameter of 650 µm are arranged in a rectangular grid.

Upstream of the first MDU, the flow channel of the piston opens up to the full diameter of the chamber (detailed view, Figure 1) to support a more homogeneous flow through the matrix and avoid dead spaces. Downstream of the second MDU, the channel narrows again. Besides, a screw cap was introduced to make the bioreactor chamber more flexible with regards to the thickness of the scaffolds inserted. Matrices with a thickness of 18 mm can be inserted into BR2 while the MDUs hold them in place. Additionally, the Luer lock connectors were implemented in the piston and housing to reduce possible leakage sites. In contrast to BR1, BR2 and the MDUs were manufactured from polyoxymethylene (POM). Materials for both bioreactors were chosen to be compatible with steam sterilization.

2.2. Incubator System

The incubator system "Incubator S 2220" was developed, manufactured, and modified by Fraunhofer IGB, Stuttgart, Germany. This custom-made system is equipped with two peristaltic pumps with a 4-channel pump head (ISM 915 and ISM 721 from ISMATEC, Wertheim, Germany) and two pressure sensors (SP 844 from MEMSCAP, Durham, NC, USA). All standard cell culture parameters like temperature, ambient CO_2, and O_2 are controlled by a Siemens SIMATIC controller. The incubator chamber is easily accessible and allows safe handling of the bioreactor parts. All parameters are monitored and controlled via integrated touch screen panel (Figure 2). The pumping speed is manually adjustable or can be controlled by a feedback loop with the pressure sensors. The incubator chamber can be either equipped with a single bioreactor system, which is equipped with two pressure sensors, an O_2-sensor, and a pinch valve for the application of hydrostatic pressure, or with up to eight separate bioreactor systems without additional sensors. Both the setups feature independent media circuits.

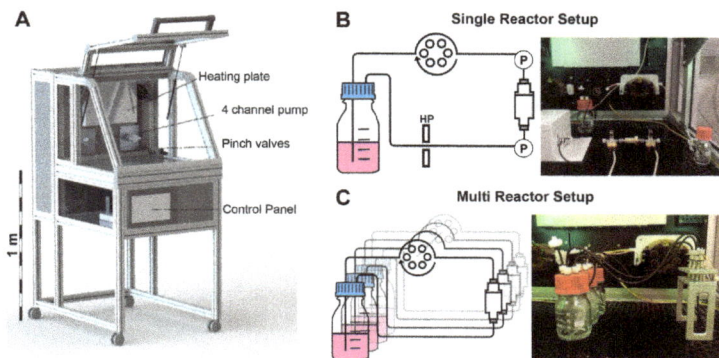

Figure 2. (**A**) Tailor made incubator system developed by Fraunhofer IGB, Stuttgart. A heating plate controls the temperature of the incubator chamber. Two 4 channel pumps can operate several tubing systems at once. A pinch valve can be used to apply hydrostatic pressure (HP). All functions are controlled via a touch screen control panel and all data can be recorded via USB port; (**B**) Single reactor setup in the incubator system: pressure sensors (P) measure the pressure differential inside the bioreactor system non-invasively; (**C**) Multi reactor setup: up to eight independent bioreactors can be operated in parallel (modified from [23], with permission from S. Karger AG, Medical and Scientific Publishers).

2.3. Pressure Sensor Characterization

The pressure in the bioreactor tubing system was studied for each sensor separately at flow rates ranging from 1.5 to 15 mL/min. For this, the pump was programmed to increase the flow rate stepwise by 0.5 mL/min every 10 min ($n = 3$). As the data acquisition system recorded each change in pressure, these measurements resulted in 4423 ± 4 data points for each flow rate. A circular bioreactor setup filled with double distilled water (37 °C) was used during sensor characterization (Figure 3).

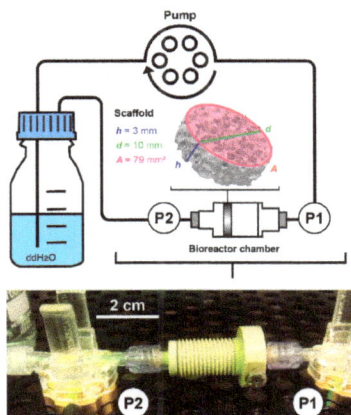

Figure 3. Scheme and picture of the bioreactor setup used during fluid shear stress prediction measurements. Water at 37 °C was pumped through the bioreactor chamber containing the porous scaffold Sponceram. Pressure sensors measured the pressure upstream (P1) and downstream (P2) of the bioreactor chamber non-invasively. Sponceram is depicted as volume rendering of a microCT scan and scaffold dimensions are given as h = height, d = diameter, and A = area.

2.4. Determination of Permeability

The permeability k is an important material constant, which allows the estimation of shear forces occurring at different flow velocities. To measure the permeability of a porous scaffold, the pressure inside the bioreactor system was measured simultaneously upstream (P_1) and downstream of the bioreactor chamber (P_2) at different flow rates (Figure 3). The differential pressure ΔP ($P_1 - P_2$) can be used together with characteristics of the biomaterial to calculate the permeability k with Darcy's law:

$$k = \frac{Q \cdot \mu \cdot h}{A \cdot \Delta P} \tag{1}$$

where Q refers to the volumetric flow rate, μ is the dynamic viscosity of water at 37 °C, h is the height, and A the area of the biomaterial. The scaffold Sponceram (Zellwerk GmbH, Eichstaedt, Germany) used in this study is a ceramic zirconium dioxide matrix (Figure 3). Porosity (66.7%) and average pore size (510 µm) were derived from µCT scans (data not shown). It was shown to have bone-like properties [25] and thus was suggested to be a suitable matrix for bone tissue engineering processes. The scaffold discs used in this study were 10 mm in diameter and 3 mm in thickness.

The differential pressure was measured at different flow rates ranging from 1.5 to 15 mL/min (increment of 0.5 mL/min) with the same setup used during sensor characterization (Figure 3). The data was recorded with the incubator's data acquisition. Each flow rate was measured for 10 min resulting in 4484 ± 71 data points for each flow rate and measurement. Three randomly picked Sponceram discs were used for the measurements (each $n = 3$).

2.5. Computational Fluid Dynamics

To estimate the flow profile and streamlines in the empty bioreactor, 3D models of both bioreactor cartridges were generated with CAD in Solidworks 2015 (Waltham, MA, USA) and imported to COMSOL Multiphysics™ (Burlington, Florence, NJ, USA). The porous media flow model was used where the steady-state Navier-Stokes equations were solved. The material was set to *water* at 37 °C, the inlet boundary condition to *velocity* with *normal inflow velocity*, and the outlet boundary condition was set to *pressure*. For all solid walls, *no slip boundary* conditions were set. The mesh was created by COMSOL with a *normal* element size. The shear stress distribution in a porous scaffold was simulated with the same model. To simplify the scaffold, a cylinder of the same dimensions with porous matrix conditions was introduced to the model. The porosity of the scaffold was set to be 66.7% and permeability was determined by measuring the pressure differential of the bioreactor in- and outlet and was set to $1.7 \pm 0.9 \times 10^{-10}$ m^2 (see Section 2.4).

A stationary study was conducted using 3.6, 17.8, and 35.6 mm/s as flow velocities at the bioreactor inlet which corresponds to the volumetric flow rates of 1.5, 7.5, and 15 mL/min. The velocity profile, streamlines, and shear stress distribution of the bioreactor chambers were plotted from these results.

2.6. Residence Time Distribution

To characterize the mixing behavior of a bioreactor system, the residence time distribution (RTD) can be obtained from wash out experiments. For this, a Dirac pulse with the tracer substance methylene blue is injected at the entrance of the bioreactor chamber. Simultaneously, the concentration of the tracer substance at the exit is measured.

First, the bioreactor volume V_R was measured by weighing the reactor chamber with and without water. The volumetric flow rate Q was measured by weighing the water that was pumped through the system within a certain time. The hydrodynamic residence time T was derived from V_R/Q.

To obtain the RTD, a Dirac pulse of 100 µL methylene blue solution (Carl Roth, Karlsruhe, Germany) 1:18 in ddH$_2$O was injected at the entrance of the bioreactor chamber during perfusion. Drops were collected 10 cm downstream in a 96 well plate and the absorbance at 688 nm of each well was measured with a plate reader (Infinite M1000, Tecan, Männedorf, Switzerland). The measurement was carried out for $t = 0\ldots 4 \times T$ at 0.6, 1.5, and 3 mL/min (at least $n = 3$). To investigate the influence of a bone like porous matrix inside the chamber, the measurements were performed with and without Sponceram.

The RTD was then derived from the data collected in the washout experiments as described before [9]. Briefly, the residence time function $E(t)$ is calculated by dividing the concentration of the tracer at each time point by the integral of the tracer concentration from 0 to $4 \times T$. To compare measurements of different bioreactors, the dimensionless residence time function $E(\Theta)$ can be derived from $E(t)$.

The tanks-in-series (TIS) model describes real bioreactors as a cascade of perfectly mixed continuous stirred tank reactors (CSTR) with N tanks in series [26]. For $N \to 1$ the bioreactor behaves like a CSTR, and for $N \to \infty$ it behaves like a plug flow reactor (PFR) without any axial mixing. Data obtained in the washout experiments was fit to the TIS model with a global curve fit using the software OriginPro (OriginLab, Northampton, MA, USA) with N as the key parameter of the following equation:

$$E(\Theta) = \frac{N(N\Theta)^{N-1}}{(N-1)!} e^{(-N\Theta)} \qquad (2)$$

A real bioreactor can also be described with the dispersion model where the dimensionless Bodenstein number Bo describes the ratio between convective transport and axial diffusion. For a system with open-open boundary conditions it can be derived from the response curve of a Dirac pulse as follows:

$$Bo = \frac{1 + \sqrt{8 \cdot \sigma_\Theta^2 + 1}}{\sigma_\Theta^2} \qquad (3)$$

$$\sigma_\Theta^2 = \frac{\sigma^2}{T^2} \qquad (4)$$

With σ^2 as the variance and σ_Θ^2 as the dimensionless variance. For $Bo \to 0$ the axial dispersion is high, indicating strong back mixing. For $Bo \to \infty$, the axial dispersion is 0, indicating no back mixing.

2.7. Fluid Shear Stress Estimation

Fluid flow induced shear stress is an important parameter in cell culture processes. Thus, after determination of the permeability, the fluid shear stress was calculated from the CFD data. For laminar flow systems, the wall shear stress τ_ω is defined by the normal velocity gradient at the wall:

$$\tau_\omega = \mu \frac{\partial u}{\partial n} \qquad (5)$$

where μ is the dynamic viscosity, u the flow velocity, and n is the x-, y-, and z-direction. Based on Equation (5), the average ($\tau_{\omega avrg}$) and maximum shear stress ($\tau_{\omega max}$) was calculated from the entire scaffold *domain* which was introduced in the COMSOL model as described in Section 2.5. Furthermore, the CFD derived shear stress was compared with a model proposed by Vossenberg et al. [27] which uses the permeability constant k as an indicator for shear stress. It also uses the permeability constant k to calculate $\tau_{\omega avrg}$ and $\tau_{\omega max}$ at a flow velocity of 100 µm/s:

$$\tau_{\omega avrg} = 9.82 \cdot 10^{-12} k^{-0.914} \qquad (6)$$

$$\tau_{\omega max} = 3.36 \cdot 10^{-10} k^{-0.807} \qquad (7)$$

In fact, Equations (5)–(7) are only valid in systems with laminar flow assuming Darcy's law is applicable. This is the case as long as the interstitial Reynolds number $Re_i < 8$ [28]. Consequently, Re_i was calculated from the bioreactor and scaffold parameters:

$$Re_i = \frac{\rho \psi u D_P}{\mu (1 - \varepsilon)} \tag{8}$$

where ρ is the density of water at 37 °C, D_P is the average pore diameter, ψ is the sphericity (for simplicity assumed to be 1), and ε is the porosity.

2.8. Cell Culture

Human ASCs used in this study were isolated from female donors (42, 48, and 52 years old) as described previously [24]. Isolation from human tissue was approved by the ethics committee of the Medical University Vienna, Austria (EK Nr. 957/2011, date: 30 January 2013). All donors gave written consent. Briefly, fat tissue obtained from abdominoplasty was minced with scissors and digested with collagenase type I (Sigma Aldrich, St. Louis, MO, USA). After several centrifugation and washing steps, the stromal vascular fracture was released in a cell culture flask and ASCs were selected by plastic adherence. The donor tissue was processed within 3–6 h after surgery.

After isolation, ASCs were cultivated in standard medium composed of MEM alpha (Thermo Fisher Scientific, Waltham, MA, USA), 0.5% gentamycin (Lonza, Basel, Switzerland), 2.5% human platelet lysate (PL BioScience, Aachen, Germany), and 1 U/mL heparin (Ratiopharm, Ulm, Germany) in a humidified incubator at 37 °C and 5% CO_2. Cells were cryo-preserved in 77.5% αMEM, 12.5% HPL, 10% DMSO (Sigma Aldrich), and 1 U/mL heparin, as described previously [9]. After thawing, the cells were expanded for two passages in T-flasks (Sarstedt, Nümbrecht, Germany) and harvested via accutase (GE healthcare, Little Chalfont, UK) treatment to be used for cultivation in the bioreactor system. For cell culture experiments, cells of the three donors were mixed in equal amounts before seeding.

2.9. Bioreactor Cultivation

To evaluate the influence of the different flow profiles and shear stress distributions of BR1 and BR2 on osteogenic differentiation, human ASCs were cultivated for 21 days on Sponceram. Prior to seeding, Sponceram matrices were steam sterilized. After steam sterilization, Sponceram matrices were seeded with 50 μL of a 6×10^6 cells/mL cell suspension of ASCs (passage 2, $n = 3$ donors) and incubated for 2 h at 37 °C before they were carefully covered with medium. Cells seeded on a 12 well plate (4000 cells/cm^2) served as the 2D static control. The seeded matrices were transferred to the bioreactor chamber (3D dynamic) after 3 days or were kept in the well of a 6 well plate (3D static). The bioreactors and 6 well plates were filled with either 10 mL osteogenic differentiation medium (ODM; standard medium supplemented with 5 mM beta-glycerolphosphate, 0.1 μM Dexamethasone, 200 μM L-ascorbate-2-phosphate, all from Sigma Aldrich, St. Louis, MI, USA) whereas the 12 well plates were filled with 2 mL. Perfusion in the bioreactors were set to a flow rate of 1.5 mL/min. Cells were cultivated for 21 days and the medium was changed every 2–3 days (1 mL for determination of ALP activity, glucose, and lactate) while 6 mL were exchanged on day 7 and 14.

2.10. DNA Quantification

Prior to DNA extraction, 3D samples were grinded whereas 2D samples were detached by accutase treatment. Lysis buffer containing 0.1 mg/mL Proteinase K (Sigma Aldrich) was then added to each sample before incubating the samples (3 h, 37 °C, 100 rpm). DNA was precipitated with 100% ethanol and after centrifugation (14,000× g, 20 min, 4 °C) was washed with 70% ethanol. The pellet was dried, resuspended in TE buffer, and stored at 4°C. DNA was quantified using the Invitrogen™

Quant-It™ PicoGreen® dsDNA Assay Kit according to instructions provided by the manufacturer (Invitrogen, Carlsbad, CA, USA).

2.11. Alkaline Phosphatase Activity

Alkaline phosphatase (ALP) as an osteogenic marker can be detected in the cell culture supernatant. To determine the ALP activity, the supernatants were transferred into a 96-well plate (8×80 μL per condition) and 20 μL of a p-nitrophenyl phosphate stock solution (Sigma Aldrich) was added to each well. After 60 min of incubation, the absorption at 405 nm was detected and the ALP activity was calculated from this data.

2.12. DAPI Staining

Prior to staining the cell nuclei with 4′,6-diamidin-2-phenylindol (DAPI), the samples were fixated with 96% ethanol. Cells or scaffolds were rinsed with PBS, covered with DAPI staining solution (1 μL DAPI stock in DAPI buffer), and incubated for 20 min at room temperature. Subsequently, the cells were rinsed with PBS twice and documented by fluorescence microscopy (excitation/suppression filter: 360/470 nm). For better comparability, the fluoresence micrographs of the different conditions were taken with consistent parameters (exposure, gain, and gamma) and images of the entire sample were acquired at 4-fold magnification (30–40 images per sample). Subesquently, images were digitally merged with the software "Microsoft Image Compositor Editor" to give a comprehensive overview of the entire scaffold.

2.13. Matrix Mineralization

Calcium content of the extracellular matrix was observed by calcein (calcium deposition) and von Kossa (phosphate deposition) stain. Samples were fixated with 96% ethanol. For calcein staining, samples were washed with ddH$_2$O and incubated over night at 4 °C in calcein staining solution (0.1 μg/mL in ddH$_2$O; Sigma Aldrich). Afterwards, cells were washed with PBS and observed with a fluorescence microscope (exposure: 205 ms; gain: 1×; gamma: 1×).

To observe phosphate deposition, the fixated cells were washed with ddH$_2$O and incubated in 1 mL AgNO$_3$ solution (5% w/v; Carl Roth) for 30 min and were light protected. After the samples were rinsed with ddH$_2$O and exposed to UV light for 2 min (each side), they were incubated in Na$_2$S$_2$O$_3$ (5% w/v, Sigma Aldrich). Samples were rinsed with ddH$_2$O and the staining was documented with a flatbed scanner.

2.14. Statistical Analysis

All data are expressed as mean values ± standard deviation. The data was analyzed using Microsoft Excel, OriginPro, and GraphPad Prism. Multiple comparisons were carried out using one-way analysis of variance followed by the Dunnett's or Tukey's multiple comparisons test. Values of $p < 0.01$ with a confidence interval of 99% were defined as statistically significant.

3. Results

3.1. Flow Profile and Residence Time Distribution

A homogeneous flow profile throughout the bioreactor chamber is preferable since nutrition supply and waste removal is crucial in every 3D cultivation. Thus, a computational study was conducted to simulate the distribution of flow velocity and streamlines in the bioreactor with and without a scaffold inserted. The flow profile of the empty BR1 indicates a higher flow velocity exclusively in the center of the chamber (Figure 4). In contrast, the outer areas display velocities close to zero while circular streamlines indicate dead spaces. After introducing the scaffold into the model, higher flow velocities were especially observed in the center region. In contrast, the medium distribution units (MDUs) of BR2 seem to support a homogeneous flow profile. The average and maximum velocity of the fluid that passed through the scaffold was derived from the CFD data (Table 1). In BR1 the maximum flow velocity is 16-fold higher than the average velocity whereas it is only 5-fold higher in BR2. Therefore, the MDUs seem to support a homogeneous flow velocity profile and reduce flow velocity peaks.

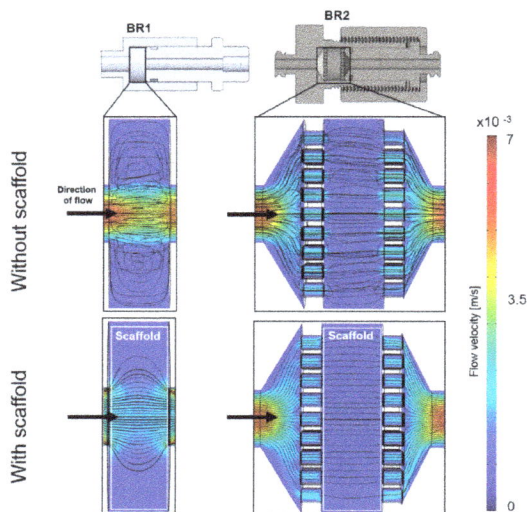

Figure 4. Computational model of the flow profile with streamlines of the bioreactor chambers BR1 and BR2 without and with a scaffold inserted (porosity: 66.7%, permeability 1.74×10^{-10} m^2) at a flow rate of 1.5 mL/min (3.6 mm/s). Streamlines and the flow profile indicate a more homogeneous flow distribution in BR2 due to the medium distribution units.

Table 1. CFD derived data of the average and maximum velocities of fluid passing the scaffold inside the bioreactor chambers BR1 and BR2.

Bioreactor	Inlet Velocity [mm/s]	Average Velocity [mm/s]	Maximum Velocity [mm/s]	Maximum/Average Velocity
	3.5	0.4	6.2	16.0
BR1	17.7	1.9	30.0	15.4
	35.4	3.9	56.5	14.5
	3.5	0.3	1.7	5.3
BR2	17.7	1.6	8.6	5.3
	35.4	3.2	17.3	5.4

The residence time distribution (RTD) of a bioreactor characterizes the mixing behavior, and consequently is an important parameter in bioprocess engineering. In this study we compared the RTDs of two different bioreactor chambers at different flow rates with or without a scaffold inserted. Ratios of the ideal hydrodynamic residence time T and the real mean residence time T_m at different flow rates were compared (Table 2, Figure 5).

T is a theoretical value that describes the time a fluid volume needs to pass through the bioreactor when no back mixing occurs (like in an ideal PFR). As expected, both bioreactors did not behave like an ideal PFR and T_m was higher than T (12–37%), indicating back mixing. Moreover, T_m/T does not appear to be affected by flow rate or bioreactor geometry significantly. However, in BR2 the ratio of T_m/T seems to decrease with higher flow rates and is lower at 3 mL/min than in BR1 showing a more PFR like behavior.

Table 2. Differences of the mean residence time *Tm* to the ideal hydrodynamic residence time *T* in percent (at least *n* = 3).

Condition	Difference to *T* [%]		
Flow rate (mL/min)	0.6	1.5	3.0
BR1 empty	16 ± 0.1	12 ± 0.4	30 ± 0.3
BR2 empty	37 ± 1.1	27 ± 0.8	23 ± 1.7
BR1 with scaffold	18 ± 0.1	31 ± 1.7	23 ± 1.7
BR2 with scaffold	27 ± 0.6	16 ± 0.5	15 ± 0.7

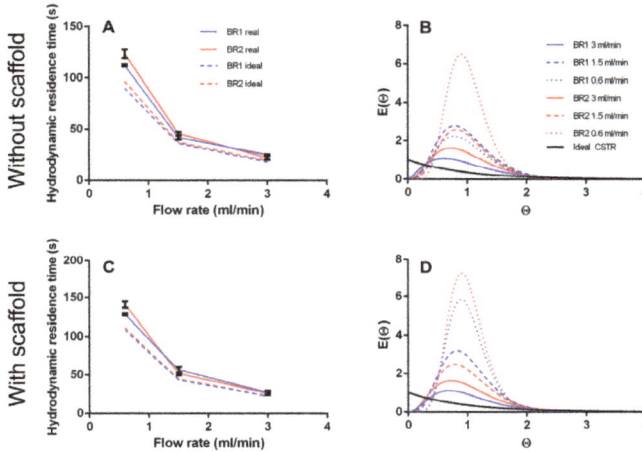

Figure 5. (**A**) Hydrodynamic residence time and (**B**) residence time distribution of BR1 and BR2 without inserted scaffold. (**C**) Hydrodynamic residence time and (**D**) residence time distribution of BR1 and BR2 with inserted scaffold.

The curve fit of the RTDs of the empty bioreactor indicate a correlation between increasing flow rate and stronger mixing, resulting in a more CSTR like mixing behavior (Figure 5). At lower flow rates the fluid behaves more like that in a PFR. Besides, inserting a scaffold seems to inhibit back mixing and instead promotes a more uniform flow.

The tanks in series and the Bodenstein number *Bo* were both derived from the RTDs and are compared in Table 3. Generally, more tanks in series were observed at lower flow rates indicating a more plug flow like behavior. The chamber geometry of BR2 seems to support a more laminar flow with less back mixing whereas the chamber of BR1 seems to improve back mixing. Again, the insertion of a scaffold into the fluid pathway supports a plug flow behavior. For the flow rates of 1.5 and 3.0 mL/min, *N* is between 2.5 and 5.7. When *N*→1 the system is considered to be mixed completely. Thus, a flow rate of 1.5 mL/min and higher can be considered as optimal mixing.

The Bodenstein number was derived from the raw data (not from the global curve fit) and should behave similar to the number of tanks-in-series model. It is highest at the lowest flow rate indicating low axial dispersion (except for BR1 empty). Although, *Bo* does not decrease with the flow rate considerably like the number of tanks in series.

The data from the computational model together with the RTDs indicate almost optimal mixing in BR1 with a heterogeneous flow profile. In contrast, BR2 demonstrated less mixing but a homogeneous flow velocity profile.

Table 3. Tanks in series and Bodenstein number of both bioreactors with and without the scaffold. Tanks in series are derived from a global curve fit (*n* = 3). The Bodenstein number was calculated from the raw data set.

Flow Rate (mL/min)	Tanks in Series			Bodenstein Number		
	0.6	1.5	3.0	0.6	1.5	3.0
BR1 empty	4.6	4.8	2.5	6.1	6.3	7.0
BR2 empty	9.7	4.8	2.5	7.4	6.7	5.9
BR1 with scaffold	9.7	5.7	3.3	8.0	7.3	7.7
BR2 with scaffold	10.7	4.7	3.4	9.1	7.2	7.6

3.2. Sensor Characterization and Determination of Permeability

The installed pressure sensors are usually used for medical purposes to monitor the blood pressure of patients in intensive care units by piezoresistive transducers. According to the manufacturer's specifications, the sensors together with the amplifier have a pressure range of −20 to 300 mmHG (−2.6–40 kPa). Indeed, flow rates used in 3D perfusion cell culture are often as low as 0.3–3 mL/min [29], resulting in a flow induced pressure of approximately <3 kPa (depending on the tubing diameter). Hence, these implemented sensors may not measure as accurately as in their original setting. Still, the characterization of the sensors show a strong linear correlation of pressure and flow rate (R^2 > 0.997, p < 0.0001; Figure 6). Interestingly, the standard deviation of the measurements decreased with increasing flow rate indicating a more accurate measurement at higher flow rates. The pressure differential of both sensors was found to be very constant at 338 ± 20 Pa (2.5 ± 0.5 mmHg).

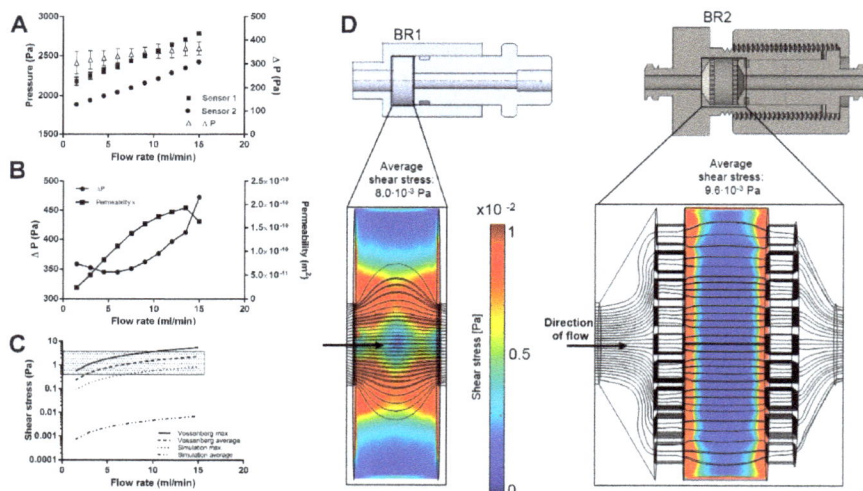

Figure 6. (**A**) Sensor characterization at different flow rates. ΔP depicts the differential of both sensors; (**B**) Determination of the permeability of the porous scaffold Sponceram. ΔP and the resulting permeability k at different flow rates; (**C**) Shear stress prediction based on the determination of the permeability of Sponceram and further calculation with the Vossenberg model or CFD data. Depending on the flow rate, the analytical model predicts shear forces in the physiological in vivo range of 0.3–3 Pa (shaded area) whereas the computational modeling predicts shear forces that are approximately three orders of magnitude lower; (**D**) Computational simulation of the shear stress conditions. BR1 and BR2 with porous scaffold (porosity: 66.7%, specific permeability 1.74×10^{-10} m²) at a flow rate of 15 mL/min. The model indicates a more homogeneous distribution of shear stress in BR2.

To determine the permeability k of Sponceram, the pressure differential upstream and downstream of the bioreactor was recorded at different flow rates. Since k is a material constant, ΔP should increase proportionately with the flow rate. Although the pressure sensors appear to be accurate, the calculated permeability changed with the flow rate (Figure 6). ΔP increased linearly for flow rates >7.5 mL/min while k remained almost stable. Therefore, the average of the calculated k-values from 9 to 15 mL/min ($k = 1.7 \pm 0.9 \times 10^{-10}$ m^2) was inserted in the shear stress equations and the computational model. Similarly, the permeability of Sponceram with a higher porosity (80%) than the Sponceram discs used in this study was found to be $k = 1.88 \times 10^{-8}$ m^2 and the permeability of cancellous bone was approximately 2.1×10^{-9} m^2 as reported before [25]. The lower permeability measured in the present study can be attributed to the lower porosity (67%) of Sponceram used in this study.

3.3. Fluid Shear Stress Estimation

Mechanical stimuli such as shear forces are commonly known to influence cellular behavior. Fluid shear forces can induce the differentiation of stem cells towards specific lineages which is important in all dynamic cell culture processes [30]. In this study we conducted a computational simulation to compare the shear stress distribution in two different bioreactor chambers on a bone like scaffold. The interstitial Reynolds number Re_i was calculated first to ensure that Darcy's law is applicable ($Re_i < 8$). Re_i was found to be between 0.28 and 7 for flow rates between 1.5 and 15 mL/min, thus indicating a laminar flow. Depending on the flow rate and bioreactor geometry, the average shear stress calculated by the simulation was between 0.1×10^{-2} and 1×10^{-2} Pa, and the maximum shear stress was between 8.8×10^{-2} and 85.2×10^{-2} Pa (Table 4). Although the average shear stress is similar in both bioreactor models, the maximum shear stress is about 2-fold higher in BR1, indicating higher stress peaks due to the bioreactor geometry.

Table 4. Data from the CFD simulation and Vossenberg model describing the average and maximum shear stress inside the scaffold.

Bioreactor	Inlet Velocity [mm/s]	CFD Simulation		Vossenberg Model	
		Average Shear Stress [10^{-2} Pa]	Maximum Shear Stress [10^{-2} Pa]	Average Shear Stress [10^{-2} Pa]	Maximum Shear Stress [10^{-2} Pa]
BR1	3.5	0.1	8.8	23.4	56.0
	17.7	0.4	43.4	117.2	280.0
	35.4	0.8	85.3	234.4	560.0
BR2	3.5	0.1	4.6	-	-
	17.7	0.5	22.9	-	-
	35.4	1.0	45.4	-	-

Generally, the results of the simulation indicate higher shear forces at the surfaces of the scaffold where the fluid enters. In BR1, higher shear forces occur at the entrance, core, and exit of the scaffold whereas the outer areas show very low shear forces. In contrast, BR2 displays a much more homogeneous shear stress distribution throughout the scaffold (Figure 6).

Furthermore, an analytical model that uses the permeability k as an indicator for wall shear stress was used to estimate shear stress and was compared to the CFD derived calculation. The Vossenberg model predicts average shear stress between 0.23 and 2.34 Pa which is approximately 3 orders of magnitude lower than that predicted by the CFD simulation. Maximum shear forces were calculated to be between 0.56 and 5.6 Pa which is approximately 6-fold higher than that estimated by the CFD simulation (Figure 6). Simplification of the scaffold in the computational model probably lowers the prediction of shear forces. Also, in a study by Jungreuthmayer, et al. [31] where µCT data of a scaffold was used for CFD analysis, the computational model underestimated shear forces compared to the analytical models.

Although estimations from computational and analytical models differ from each other, the shear forces calculated from the analytical models were in the range of in vivo shear stress of bone, which is expected to be between 0.3 and 3 Pa [32]. Therefore, this bioreactor-incubator system, together with this scaffold, may represent a suitable system for bone tissue engineering.

3.4. Bioreactor Cultivation

To investigate the influence of the different shear stress distributions and flow profiles of BR1 and BR2 on osteogenic differentiation of human ASCs, cells were cultivated for 21 days in ODM on the 3D ceramic matrix Sponceram. Seeded matrices in 6 well plates (3D static) and cells seeded in conventional 12 well plates (2D static) served as controls. The cell number of each sample was determined indirectly via DNA quantification and was highest in the 3D and 2D control group. However, twice as many cells were found in BR2 compared to BR1 where the cell number did not change significantly compared to day 0 (Figure 7). These findings were also confirmed by DAPI staining (Figure 8).

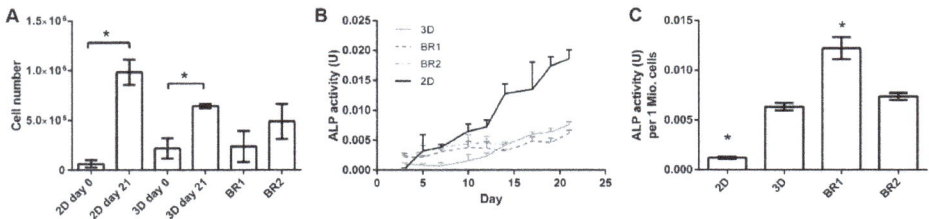

Figure 7. (**A**) Cell numbers, (**B**) course of alkaline phosphatase (ALP) activity and (**C**) ALP activity per cell of adipose derived mesenchymal stem cells (ASCs) cultivated on Sponceram in a perfusion bioreactor or under 3D and 2D static conditions. Data are represented as mean ± SD ($n = 3$); * significant difference of the indicated conditions in panel (**A**) or to 3D static in panel (**C**) with a confidence interval of 99% and $p < 0.001$.

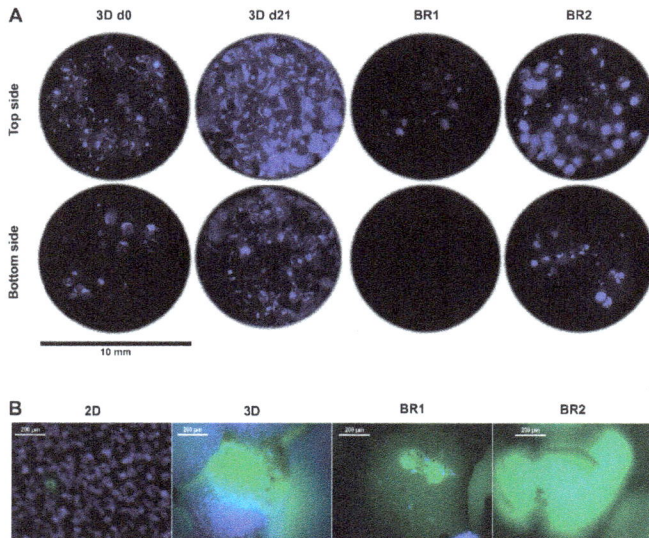

Figure 8. (**A**) 4′,6-diamidin-2-phenylindol (DAPI) stain and (**B**) DAPI-calcein double stain of Sponceram after 21 days of cultivation with ASCs in a perfusion bioreactor or under static conditions.

Glucose consumption and lactate production were steady throughout the entire cultivation period although lactate production decreased slightly after 12 days in dynamic conditions (Figure 9). The overall glucose consumption and lactate production was found to be highest in 3D static conditions. However, consumption and production per cell between day 19 and 21 indicates a higher glycolytic activity in cells cultivated in dynamic conditions (Table 5).

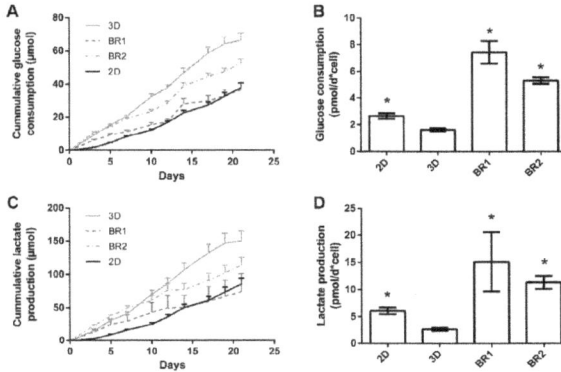

Figure 9. Course of (**A**) glucose consumption and (**C**) lactate production, (**B**) glucose consumption and (**D**) lactate production per cell between day 19 and 21 of ASCs cultivated on Sponceram in a perfusion bioreactor or under 3D and 2D static conditions. Data are represented as mean $\pm$ SD ($n = 3$); * indicates significant difference to 3D static with a confidence interval of 99% and $p < 0.001$.

Table 5. Glucose consumption and lactate production of ASCs under different cultivation conditions ($n = 3$).

Condition	Cumulative Glucose Consumption (µmol)	Cumulative Lactate Consumption (µmol)	Ratio
2D	37.9 ± 2.9	85.5 ± 8.7	2.26
3D	66.3 ± 4.5	150.1 ± 15.0	2.26
BR1	36.6 ± 4.1	74.2 ± 26.7	2.15
BR2	52.8 ± 2.4	113.6 ± 11.7	2.02

ALP activity, a marker for osteogenic differentiation, increased after approximately 12–14 days in 2D and 3D static conditions. In contrast, in dynamic conditions it was elevated from day 3 on but did not increase as much as under static conditions after 14 days. However, ALP activity per cell revealed an increased activity in all 3D conditions compared to 2D conditions, while the activity per cell in BR1 was found to be higher compared to BR2 (Figure 7).

Matrix mineralization was determined by calcein and von Kossa stain. Calcium depositions were found in all conditions (Figures 8 and 10). However, only few stained areas were observed. Furthermore, phosphate depositions were found to be increased in BR1 but only slightly in BR2 and 3D static conditions.

Figure 10. Von Kossa stain of Sponceram after 21 days of cultivation with ASCs in a perfusion bioreactor or under static conditions.

4. Discussion

In this study we developed a miniaturized perfusion bioreactor system together with a specialized incubator system for use in different cell culture applications and for bioprocess optimization. At first, the incubator system was developed and modified to serve as a flexible platform for bioreactor cultivation. For this, it was equipped with pressure sensors, hydrostatic pressure valves, and multi-channel pumps that can be programmed according to the user's needs.

Next, in order to establish a flexible perfusion system we designed, manufactured, characterized, and compared two different bioreactor chambers. The general advantages of both chambers are depicted in Table 6. Hence, the flow velocity profiles of both the bioreactors were simulated and the mixing behavior was characterized by measuring the residence time distribution. Though, BR2 displayed a more PFR like mixing behavior compared to BR1, flow rates of 1.5 mL/min and higher were considered to provide sufficient mixing. Although the RTDs from both bioreactors did not differ considerably, the flow field experienced by cells might be different [33]. Recirculation areas found in the CFD simulation of BR1 indicate a different flow field than in BR2 where MDUs prevent those recirculation areas. The tanks-in-series model also revealed less axial dispersion in BR2. Indeed, MDUs in BR2 promote a uniform flow velocity profile and plug flow like behavior throughout the scaffold while preserving proper mixing.

Table 6. Conclusive overview on the characteristics of BR1 and BR2.

Aspect	BR1	BR2
Advantages	- Sufficient mixing - Increased matrix mineralization - Increased ALP activity/cell	- Sufficient mixing - Homogeneous flow profile and shear stress distribution throughout the scaffold - Increased proliferation - More homogeneous growth on the scaffold
Disadvantages	- Inhomogeneous flow profile and shear stress distribution	- Comparably low matrix mineralization

Furthermore, the permeability of a porous scaffold was determined with pressure sensors in order to estimate shear stress with the model proposed by Vossenberg and a CFD simulation. Medium distribution units of BR2 proved to be beneficial in promoting a uniform flow velocity and shear stress distribution throughout the scaffold. A homogeneous flow and shear stress profile is highly preferable since this ensures maximum control and reproducible results during the experiments. To investigate effects of the different flow and shear stress profiles, both bioreactor chambers were compared in a proof-of-concept bioreactor cultivation study.

For this, the flow rate was chosen according to the calculations of the simulation and mathematical model as follows. The average shear stress at a flow rate of 1.5 mL/min was between 0.001 (CFD) and 0.23 Pa (Vossenberg) and the maximum shear stress was between 0.08 and 0.56 Pa which only partially lies in the physiologic in vivo shear stress of 0.3–3 Pa. However, osteogenic differentiation performed in 3D conditions was found to be increased with shear stress by about one order of magnitude lower than the in vivo shear stress [6]. Seemingly, lower shear forces are sufficient to induce osteogenic differentiation. Also, the maximum shear stress was shown to increase dramatically compared to the average shear stress with increasing flow rate if the permeability constant k is below $1 \cdot \times 10^{-10}$ m^2 [27]. In the present work, the permeability constant k of the scaffold was found to be $1.7 \pm 0.9 \times 10^{-10}$ m^2. Furthermore, the Vossenberg model was developed for scaffolds with perpendicular fibers and thus the permeability might not be as predictive for shear stress as in a porous scaffold. Also, the CFD simulation lacks accuracy since it is based on a simplified geometry of the scaffold. Taken all together, the flow rate of the proof of concept study was set to 1.5 mL/min to generate shear forces at the lower end of the physiologic range in order to avoid excessive washout of cells.

After 21 days of cultivation, the cell number was found to be increased in BR2 compared to BR1 while the cell distribution was more homogeneous as indicated by DAPI staining. However, cell number and glucose consumption indicate lower proliferation in dynamic conditions compared to static conditions. Thus, cells in dynamic conditions might have been subject to washout by perfusion. The ratio of lactate production per mole consumed glucose is commonly used as an index for anaerobic metabolism which occurs mainly in proliferating mesenchymal stem cells. These cells display a higher ratio since they generate energy rather by anaerobic glycolysis than by oxidative phosphorylation [34]. Indeed, cells cultivated under dynamic conditions display a lower ratio (2.02–2.15) than under static conditions (2.26) which might indicate a shift to oxidative phosphorylation and therefore to differentiation. Also, ALP activity per cell was found to be higher in dynamic conditions and phosphate deposition was only visible in BR1. These findings suggest increased differentiation in dynamic conditions. However, the glucose consumption and lactated production per cell was found to be significantly higher in dynamic conditions compared to static conditions. Controversially, a previous study by Pattappa et al. [35] reported reduced glycolytic activity during osteogenic differentiation. However, this study was carried out in conventional 2D static conditions. In contrast, 3D dynamic conditions increase the mass transfer of nutrients, oxygen, and waste products, which together with mechanical stimulation might alter metabolic activity in comparison to 2D static cultivation conditions.

Regarding the influence of the flow and shear stress profile of BR1 and BR2 on the osteogenic differentiation, cells of BR1 showed increased ALP activity per cell and matrix mineralization compared to BR2. Interestingly, phosphate depositions were present only in the center of BR1 where flow velocity and shear stress is the highest as indicated by the CFD simulations. Furthermore, the maximum flow velocity at 1.5 mL was found to be 3.6-fold higher in BR1. Although it did not affect the average shear stress, it caused a 1.9-fold increase in the maximum shear stress. Also, the maximum shear stress in BR1 was 88-fold higher than the average shear stress while it was only 46-fold higher in BR2. The findings of the CFD simulation together with the data from bioreactor cultivation suggest that in order to generate shear stress as high as in BR1, the volumetric flow rate needs to be increased in BR2. In this study, where both bioreactor chambers were operated at the same flow rate, BR1 supported an increased osteogenic differentiation of ASCs while BR2 maintained homogeneous cell growth rather than differentiation. Since it is likely that differentiation was induced by higher shear stress in the center of BR1, a higher flow rate resulting in similar shear stress conditions might foster comparable differentiation in BR2 as well. In conclusion, depending on the mathematical model, shear stress calculations derived from characterization of the matrix and CFD simulation suggested physiologic shear stress conditions for a broad flow rate spectrum of 1.5 to 15 mL/min. However, simulations were predictable for the relative comparison of BR1 and BR2, and an optimal flow rate for the generation of physiologic shear stress still needs to be evaluated by experiments. The successful screening of different shear stress and hydrostatic pressure conditions in this system was demonstrated before on a decellularized bone matrix [23].

Perfusion systems have been proven to be beneficial for tissue engineering purposes such as bone or cartilage engineering [36,37]. For instance, numerous studies show a higher matrix deposition or upregulation of relevant genes in comparison to static culture conditions when bone precursor cells are exposed to fluid shear stress [38–43]. Although several tailor-made [10,11,44] and commercially available [45–47] perfusion systems were developed and successfully used, there is a lack of automated sensor-controlled systems that allow the cultivation of multiple independent replicates under different conditions.

However, the presented system allows the determination of the permeability of scaffolds, thus enabling estimation and control of shear stress during cultivation. Since different flow and pressure regimes can be programmed with the built-in control unit, the incubator allows comprehensive control over the mechanical environment. The bioreactor chamber developed in this study has been used with hard scaffolds (e.g., ceramics) but was designed to host also soft scaffolds (e.g., hydrogels) of different sizes via integrated grids. Together with the possibility to cultivate up to sixteen independent reactors (with an upgrade of the pumping head) in one incubator, we hypothesize it can be used not only in the field of bone tissue engineering but also in a wider variety of 3D cell culture processes. Further studies will focus on demonstrating its benefit in the testing and optimization of different dynamic cell culture conditions (including hydrostatic pressure) on other biomaterials.

5. Conclusions

A parallelizable, miniaturized perfusion bioreactor system, together with a tailor-made incubator, was developed and characterized. In particular, it was designed to be flexible and modular in terms of exchanging different parts such as the tubing, pumping head, and scaffold or adding additional instrumentation, such as pinch valves for the application of hydrostatic pressure. Integrated pressure sensors allow the estimation of fluid shear stress that cells experience on a scaffold and as a result permit the screening of the effects of different mechanical culture conditions. The effect of different flow and shear stress profiles were investigated in a proof of concept study. CFD data and shear stress calculations were found to be predictable for a relative comparison of two bioreactor geometries, but not for the prediction of the optimal flow rate. We assume a parallelizable miniaturized system where only small amounts of cells, culture medium, and scaffolds are required, that will support throughput and reproducibility, and thus will be beneficial for the optimization of dynamic 3D cell culture applications.

Acknowledgments: This work was partially supported by the FFG (Research Promotion Agency, project number 846065), Austria. The authors are thankful to the mechanical workshop of the Institute of Technical Chemistry of the Leibniz University of Hannover, Germany and to TecSense, Grambach, Austria for the manufacturing of bioreactor parts.

Author Contributions: Dominik Egger designed and performed the experiments, data analysis, and prepared the manuscript. Monica Fischer performed experiments and data analysis. Andreas Clementi performed experiments. Volker Ribitsch contributed to the design and manufacturing of the bioreactor. Jan Hansmann developed the incubator system and contributed to design of the study and preparation of the manuscript. Cornelia Kasper designed the study and contributed to the preparation of the manuscript.

Conflicts of Interest: The authors declare no conflict of interest.

References

1. Volkmer, E.; Drosse, I.; Otto, S.; Stangelmayer, A.; Stengele, M.; Kallukalam, B.C.; Mutschler, W.; Schieker, M. Hypoxia in static and dynamic 3D culture systems for tissue engineering of bone. *Tissue Eng. Part A* **2008**, *14*, 1331–1340. [CrossRef] [PubMed]
2. Goldstein, A.S.; Juarez, T.M.; Helmke, C.D.; Gustin, M.C.; Mikos, A.G. Effect of convection on osteoblastic cell growth and function in biodegradable polymer foam scaffolds. *Biomaterials* **2001**, *22*, 1279–1288. [CrossRef]

3. Stiehler, M.; Bünger, C.; Baatrup, A.; Lind, M.; Kassem, M.; Mygind, T. Effect of dynamic 3-D culture on proliferation, distribution, and osteogenic differentiation of human mesenchymal stem cells. *J. Biomed. Mater. Res. Part A* **2009**, *89*, 96–107. [CrossRef] [PubMed]

4. Wang, T.; Wu, H.; Wang, H.; Lin, F.; Sun, J. Regulation of adult human mesenchymal stem cells into osteogenic and chondrogenic lineages by different bioreactor systems. *J. Biomed. Mater. Res. Part A* **2009**, *88*, 935–946. [CrossRef]

5. Gooch, K.J.; Kwon, J.H.; Blunk, T.; Langer, R.; Freed, L.E.; Vunjak-Novakovic, G. Effects of mixing intensity on tissue-engineered cartilage. *Biotechnol. Bioeng.* **2001**, *72*, 402–407. [CrossRef]

6. Sikavitsas, V.I.; Bancroft, G.N.; Mikos, A.G. Formation of three-dimensional cell/polymer constructs for bone tissue engineering in a spinner flask and a rotating wall vessel bioreactor. *J. Biomed. Mater. Res.* **2002**, *62*, 136–148. [CrossRef] [PubMed]

7. Begley, C.M.; Kleis, S.J. The fluid dynamic and shear environment in the NASA/JSC rotating-wall perfused-vessel bioreactor. *Biotechnol. Bioeng.* **2000**, *70*, 32–40. [CrossRef]

8. Diederichs, S.; Röker, S.; Marten, D.; Peterbauer, A.; Scheper, T.; van Griensven, M.; Kasper, C. Dynamic cultivation of human mesenchymal stem cells in a rotating bed bioreactor system based on the Z® RP platform. *Biotechnol. Prog.* **2009**, *25*, 1762–1771. [CrossRef] [PubMed]

9. Neumann, A.; Lavrentieva, A.; Heilkenbrinker, A.; Loenne, M.; Kasper, C. Characterization and Application of a Disposable Rotating Bed Bioreactor for Mesenchymal Stem Cell Expansion. *Bioengineering* **2014**, *1*, 231–245. [CrossRef]

10. Bancroft, G.N.; Sikavitsas, V.I.; Mikos, A.G. Technical note: Design of a flow perfusion bioreactor system for bone tissue-engineering applications. *Tissue Eng.* **2003**, *9*, 549–554. [CrossRef] [PubMed]

11. Sailon, A.M.; Allori, A.C.; Davidson, E.H.; Reformat, D.D.; Allen, R.J.; Warren, S.M. A novel flow-perfusion bioreactor supports 3D dynamic cell culture. *Biomed. Res. Int.* **2009**, *2009*, 873816. [CrossRef] [PubMed]

12. Piola, M.; Soncini, M.; Cantini, M.; Sadr, N.; Ferrario, G.; Fiore, G.B. Design and Functional Testing of a Multichamber Perfusion Platform for Three-Dimensional Scaffolds. *Sci. World J.* **2013**, *2013*, 123974. [CrossRef] [PubMed]

13. Hansmann, J.; Groeber, F.; Kahlig, A.; Kleinhans, C.; Walles, H. Bioreactors in tissue engineering—Principles, applications and commercial constraints. *Biotechnol. J.* **2013**, *8*, 298–307. [CrossRef] [PubMed]

14. Lee, P.J.; Hung, P.J.; Rao, V.M.; Lee, L.P. Nanoliter scale microbioreactor array for quantitative cell biology. *Biotechnol. Bioeng.* **2006**, *94*, 5–14. [CrossRef] [PubMed]

15. Yu, W.; Qu, H.; Hu, G.; Zhang, Q.; Song, K.; Guan, H.; Liu, T. A microfluidic-based multi-shear device for investigating the effects of low fluid-induced stresses on osteoblasts. *PLoS ONE* **2014**, *9*, e89966. [CrossRef] [PubMed]

16. Zhao, F.; Chella, R.; Ma, T. Effects of shear stress on 3-D human mesenchymal stem cell construct development in a perfusion bioreactor system: Experiments and hydrodynamic modeling. *Biotechnol. Bioeng.* **2007**, *96*, 584–595. [CrossRef] [PubMed]

17. Jeong, J.Y.; Park, S.H.; Shin, J.W.; Kang, Y.G.; Han, K.-H.; Shin, J.-W. Effects of intermittent hydrostatic pressure magnitude on the chondrogenesis of MSCs without biochemical agents under 3D co-culture. *J. Mater. Sci. Mater. Med.* **2012**, *23*, 2773–2781. [CrossRef] [PubMed]

18. Fröhlich, M.; Grayson, W.L.; Marolt, D.; Gimble, J.M.; Kregar-Velikonja, N.; Vunjak-Novakovic, G. Bone grafts engineered from human adipose-derived stem cells in perfusion bioreactor culture. *Tissue Eng. Part A* **2009**, *16*, 179–189. [CrossRef] [PubMed]

19. Grayson, W.L.; Bhumiratana, S.; Cannizzaro, C.; Chao, P.-H.G.; Lennon, D.P.; Caplan, A.I.; Vunjak-Novakovic, G. Effects of initial seeding density and fluid perfusion rate on formation of tissue-engineered bone. *Tissue Eng. Part A* **2008**, *14*, 1809–1820. [CrossRef] [PubMed]

20. Wendt, D.; Marsano, A.; Jakob, M.; Heberer, M.; Martin, I. Oscillating perfusion of cell suspensions through three-dimensional scaffolds enhances cell seeding efficiency and uniformity. *Biotechnol. Bioeng.* **2003**, *84*, 205–214. [CrossRef] [PubMed]

21. Zhao, F.; Ma, T. Perfusion bioreactor system for human mesenchymal stem cell tissue engineering: Dynamic cell seeding and construct development. *Biotechnol. Bioeng.* **2005**, *91*, 482–493. [CrossRef] [PubMed]

22. Huang, C.; Ogawa, R. Effect of hydrostatic pressure on bone regeneration using human mesenchymal stem cells. *Tissue Eng. Part A* **2012**, *18*, 2106–2113. [CrossRef] [PubMed]

23. Egger, D.; Spitz, S.; Fischer, M.; Handschuh, S.; Glösmann, M.; Friemert, B.; Egerbacher, M.; Kasper, C. Application of a Parallelizable Perfusion Bioreactor for Physiologic 3D Cell Culture. *Cells Tissues Organs* **2017**, *203*, 316–326. [CrossRef] [PubMed]

24. Gugerell, A.; Neumann, A.; Kober, J.; Tammaro, L.; Hoch, E.; Schnabelrauch, M.; Kamolz, L.; Kasper, C.; Keck, M. Adipose-derived stem cells cultivated on electrospun l-lactide/glycolide copolymer fleece and gelatin hydrogels under flow conditions–aiming physiological reality in hypodermis tissue engineering. *Burns* **2015**, *41*, 163–171. [CrossRef] [PubMed]

25. Sanz-Herrera, J.A.; Kasper, C.; van Griensven, M.; Garcia-Aznar, J.M.; Ochoa, I.; Doblare, M. Mechanical and flow characterization of Sponceram carriers: Evaluation by homogenization theory and experimental validation. *J. Biomed. Mater. Res. B Appl. Biomater.* **2008**, *87*, 42–48. [CrossRef] [PubMed]

26. Fogler, H.S. *Elements of Chemical Reaction Engineering*; Prentice-Hall PTR: Upper Saddle River, NJ, USA, 1999.

27. Vossenberg, P.; Higuera, G.A.; van Straten, G.; van Blitterswijk, C.A.; van Boxtel, A.J.B. Darcian permeability constant as indicator for shear stresses in regular scaffold systems for tissue engineering. *Biomech. Model. Mechanobiol.* **2009**, *8*, 499–507. [CrossRef] [PubMed]

28. Chor, M.V.; Li, W. A permeability measurement system for tissue engineering scaffolds. *Meas. Sci. Technol.* **2006**, *18*, 208. [CrossRef]

29. Steward, A.J.; Kelly, D.J. Mechanical regulation of mesenchymal stem cell differentiation. *J. Anat.* **2015**, *227*, 717–731. [CrossRef] [PubMed]

30. Martin, I.; Wendt, D.; Heberer, M. The role of bioreactors in tissue engineering. *TRENDS Biotechnol.* **2004**, *22*, 80–86. [CrossRef] [PubMed]

31. Jungreuthmayer, C.; Donahue, S.W.; Jaasma, M.J.; Al-Munajjed, A.A.; Zanghellini, J.; Kelly, D.J.; O'Brien, F.J. A comparative study of shear stresses in collagen-glycosaminoglycan and calcium phosphate scaffolds in bone tissue-engineering bioreactors. *Tissue Eng. Part A* **2008**, *15*, 1141–1149. [CrossRef] [PubMed]

32. Fritton, S.P.; Weinbaum, S. Fluid and Solute Transport in Bone: Flow-Induced Mechanotransduction. *Annu. Rev. Fluid Mech.* **2009**, *41*, 347–374. [CrossRef] [PubMed]

33. Bilgen, B.; Chang-Mateu, I.M.; Barabino, G.A. Characterization of mixing in a novel wavy-walled bioreactor for tissue engineering. *Biotechnol. Bioeng.* **2005**, *92*, 907–919. [CrossRef] [PubMed]

34. Chen, C.; Shih, Y.V.; Kuo, T.K.; Lee, O.K.; Wei, Y. Coordinated changes of mitochondrial biogenesis and antioxidant enzymes during osteogenic differentiation of human mesenchymal stem cells. *Stem Cells* **2008**, *26*, 960–968. [CrossRef] [PubMed]

35. Pattappa, G.; Heywood, H.K.; De Bruijn, J.D.; Lee, D.A. The metabolism of human mesenchymal stem cells during proliferation and differentiation. *J. Cell. Physiol.* **2011**, *226*, 2562–2570. [CrossRef] [PubMed]

36. Bjerre, L.; Bünger, C.E.; Kassem, M.; Mygind, T. Flow perfusion culture of human mesenchymal stem cells on silicate-substituted tricalcium phosphate scaffolds. *Biomaterials* **2008**, *29*, 2616–2627. [CrossRef] [PubMed]

37. Janssen, F.W.; Oostra, J.; van Oorschot, A.; van Blitterswijk, C.A. A perfusion bioreactor system capable of producing clinically relevant volumes of tissue-engineered bone: In vivo bone formation showing proof of concept. *Biomaterials* **2006**, *27*, 315–323. [CrossRef] [PubMed]

38. Grellier, M.; Bareille, R.; Bourget, C.; Amédée, J. Responsiveness of human bone marrow stromal cells to shear stress. *J. Tissue Eng. Regen Med.* **2009**, *3*, 302–309. [CrossRef] [PubMed]

39. Li, Y.J.; Batra, N.N.; You, L.D.; Meier, S.C.; Coe, I.A.; Yellowley, C.E.; Jacobs, C.R. Oscillatory fluid flow affects human marrow stromal cell proliferation and differentiation. *J. Orthop. Res.* **2004**, *22*, 1283–1289. [CrossRef] [PubMed]

40. Haeri, S.M.J.; Sadeghi, Y.; Salehi, M.; Farahani, R.M.; Mohsen, N. Osteogenic differentiation of human adipose-derived mesenchymal stem cells on gum tragacanth hydrogel. *Biologicals* **2016**, *44*, 123–128. [CrossRef] [PubMed]

41. Jaasma, M.J.; O'Brien, F.J. Mechanical stimulation of osteoblasts using steady and dynamic fluid flow. *Tissue Eng. Part A* **2008**, *14*, 1213–1223. [CrossRef] [PubMed]

42. Bancroft, G.N.; Sikavitsas, V.I.; Van Den Dolder, J.; Sheffield, T.L.; Ambrose, C.G.; Jansen, J.A.; Mikos, A.G. Fluid flow increases mineralized matrix deposition in 3D perfusion culture of marrow stromal osteoblasts in a dose-dependent manner. *Proc. Natl. Acad. Sci.USA* **2002**, *99*, 12600–12605. [CrossRef] [PubMed]

43. Scaglione, S.; Wendt, D.; Miggino, S.; Papadimitropoulos, A.; Fato, M.; Quarto, R.; Martin, I. Effects of fluid flow and calcium phosphate coating on human bone marrow stromal cells cultured in a defined 2D model system. *J. Biomed. Mater. Res. Part A* **2008**, *86*, 411–419. [CrossRef] [PubMed]

44. Gardel, L.S.; Correia-Gomes, C.; Serra, L.A.; Gomes, M.E.; Reis, R.L. A novel bidirectional continuous perfusion bioreactor for the culture of large-sized bone tissue-engineered constructs. *J. Biomed. Mater. Res. Part B Appl. Biomater.* **2013**, *101*, 1377–1386. [CrossRef] [PubMed]

45. Hernández-Córdova, R.; Mathew, D.A.; Balint, R.; Carrillo-Escalante, H.J.; Cervantes-Uc, J.M.; Hidalgo-Bastida, L.A.; Hernández-Sánchez, F. Indirect three-dimensional printing: A method for fabricating polyurethane-urea based cardiac scaffolds. *J. Biomed. Mater. Res. Part A* **2016**, *104*, 1912–1921. [CrossRef] [PubMed]

46. Porter, B.D.; Lin, A.S.P.; Peister, A.; Hutmacher, D.; Guldberg, R.E. Noninvasive image analysis of 3D construct mineralization in a perfusion bioreactor. *Biomaterials* **2007**, *28*, 2525–2533. [CrossRef] [PubMed]

47. Filipowska, J.; Reilly, G.C.; Osyczka, A.M. A single short session of media perfusion induces osteogenesis in hBMSCs cultured in porous scaffolds, dependent on cell differentiation stage. *Biotechnol. Bioeng.* **2016**, *113*, 1814–1824. [CrossRef] [PubMed]

bioengineering

MDPI

Article

Multivariate Curve Resolution and Carbon Balance Constraint to Unravel FTIR Spectra from Fed-Batch Fermentation Samples

Dennis Vier [1,*], Stefan Wambach [2], Volker Schünemann [1] and Klaus-Uwe Gollmer [3]

[1] AG Biophysik und Medizinische Physik, Technische Universität Kaiserslautern, Kaiserslautern 67663, Germany; schuene@physik.uni-kl.de

[2] Fachbereich Bioverfahrenstechnik, Hochschule Trier, Umwelt-Campus Birkenfeld, Birkenfeld 55761, Germany; s11a81@umwelt-campus.de

[3] Fachbereich Angewandte Informatik, Hochschule Trier, Umwelt-Campus Birkenfeld, Birkenfeld 55761, Germany; k.gollmer@umwelt-campus.de

* Correspondence: dennisvier@gmail.com; Tel.: +49-170-2056148

Academic Editor: Christoph Herwig
Received: 30 October 2016; Accepted: 19 January 2017; Published: 25 January 2017

Abstract: The current work investigates the capability of a tailored multivariate curve resolution–alternating least squares (MCR-ALS) algorithm to analyse glucose, phosphate, ammonium and acetate dynamics simultaneously in an *E. coli BL21* fed-batch fermentation. The high-cell-density (HCDC) process is monitored by ex situ online attenuated total reflection (ATR) Fourier transform infrared (FTIR) spectroscopy and several in situ online process sensors. This approach efficiently utilises automatically generated process data to reduce the time and cost consuming reference measurement effort for multivariate calibration. To determine metabolite concentrations with accuracies between ± 0.19 and $\pm 0.96 \cdot gL^{-1}$, the presented utilisation needs primarily—besides online sensor measurements—single FTIR measurements for each of the components of interest. The ambiguities in alternating least squares solutions for concentration estimation are reduced by the insertion of analytical process knowledge primarily in the form of elementary carbon mass balances. Thus, in this way, the established idea of mass balance constraints in MCR combines with the consistency check of measured data by carbon balances, as commonly applied in bioprocess engineering. The constraints are calculated based on online process data and theoretical assumptions. This increased calculation effort is able to replace, to a large extent, the need for manually conducted quantitative chemical analysis, leads to good estimations of concentration profiles and a better process understanding.

Keywords: multivariate curve resolution; *E. coli*; fed-batch; fermentation; carbon mass balance constraint; soft constraints; alternating least squares; hybrid modelling

1. Introduction

Multivariate curve resolution (MCR) with constrained alternating least squares (ALS), as described by Tauler et al. [1], is a powerful method to deconvolve overlapping spectral signals from chemical and biological reaction systems. The intended purpose is commonly the estimation of concentrations of individual components $\mathbf{C}$ or the identification of unknown spectral profiles $\mathbf{S}$ in complex aqueous solutions; generally, MCR has the ability to estimate both simultaneously from a data matrix $\mathbf{X}$. The specific feature of MCR is the decomposition of $\mathbf{X}$ in a physically or chemically meaningful way. Besides MCR and ALS, other bilinear modelling methods and different algorithms can be utilised for the decomposition of $\mathbf{X}$, with various resolution performances and limitations [2]. Because of its flexibility and popularity [3], MCR with the ALS algorithm is used in this work. A tutorial for the application of MCR to analyse multicomponent systems, with special focus on the ALS algorithm, is given in [4]. The main aspects to be considered—such as data set configurations, initial estimates and applicable constraints—are described. A central issue of bilinear decomposition is the impact which particular constraints, initial estimates and the applied algorithm may have on the uniqueness of solutions in the presence of rotational ambiguities [5]. In the present work, the initial MCR settings and constraints for the analysis of FTIR-spectra from an *E. coli* fed-batch bioprocess are described in detail.

Compared with other established chemometric analysis methods, MCR has the potential of simultaneous resolution and quantitation of all mixture components without their chemical or physical separation [5,6]. Besides the recovery of qualitative and quantitative information about analytes, the identification of unknown interferents is possible [7]. In comparison to other multivariate calibration methods, the calibration effort of MCR can be decreased significantly by application of appropriate constraints [8]. Given a suitable set of constraints, this paper demonstrates that single measurements of pure solved analytes suffice to perform quantitative MCR analysis of respective fermentation data.

MCR-ALS has been employed for years in different research fields, especially in chemical reaction processes monitored by different spectroscopic techniques such as X-ray absorption [9], fluorescence, nuclear magnetic resonance, Raman, Near-Infrared and FTIR [10]. In biochemical and biophysical processes, MCR was used to analyse protein and nucleic acids systems concerning denaturation processes, protonation equilibria or complexation processes [11]. Among other chemometric methods, MCR was utilised in single biological cell analysis to unmix information from hyperspectral images [12]. In reference to fermentation processes, several applications of MCR-ALS have been published, e.g., in monitoring alcoholic fermentations with *S. cerevisiae* [13,14], milk lactic acid fermentations with *Streptococcus* and *Lactobacillus* strains [15] and the quantification of penicillin V in bioprocesses with *Pencillium chrysogenum* [16]. In this work, MCR-ALS with tailored constraints is applied to estimate metabolism-relevant concentrations in high cell density cultivation (HCDC) fed-batch processes with *E. coli* BL21 (DE3) pET28a. The cultivated organism produces the recombinant and pharmaceutically utilisable enzyme cytochrome p450 after induction. For evaluating process kinetics and optimising the growth of microorganisms, it is useful to obtain estimations about the quantitative changes of carbon, nitrogen and phosphate sources as well as of metabolic products such as acetic acid in the fermentation broth over process runtime. The aim of this study is the resolution of these substances by a tailored MCR-ALS algorithm.

To monitor the composition of fermentation media, ATR-FTIR spectroscopy is employed as an in-line ex situ analyser. Spectral information from the fermentation process is provided online by automatic cell-free sampling of fermentation broth through an ATR flow-cell. Because of the continuous sterile sampling, the constraint of invariance of the total concentration (closure) is applicable, as described below. An automated in-line flow system can cause the problem of CO_2 and air bubbles, as well as biofilms on the ATR surface [17]. Therefore, biofilms are inhibited by employing cell-free sampling and initial ethanol-cleaning of the flow-system. As technical gas bubble prevention, just gas-tight polytetrafluoroethylene (PTFE) tubes are implemented. However, principally, the problem of gas bubbles can be handled mathematically by the MCR algorithm as shown in this study.

FTIR spectroscopy is an established technique in bioprocess monitoring and, as other IR techniques such as near-infrared (NIR) and Raman, it combines the advantages of non-invasiveness and fast simultaneous measurement of multiple solved substances [18]. An overview of advantages and disadvantages of different spectroscopy techniques in bioreactor monitoring has been described [19]. In the mid-infrared region, covered by FTIR, most excitations of fundamental molecular vibrations can be found. Especially the fingerprint area (1500–500 cm^{-1}) exhibits specific patterns of media compounds [20]. By contrast, the peaks in the NIR spectrum consist of overtones and combinations from primary MIR signals and are less distinctive. In comparison to NIR, the MIR region exhibits a higher selectivity thus allowing for a better detection of overlapping component spectra in complex aqueous mixtures. Raman spectroscopy is not sensitive to water and the small peak widths of solved components are main advantages of this technique in bioprocess monitoring [21]. However, the Raman scattering is generally weaker than the FTIR signal, while higher concentrations of target analytes are required. A comparison of FTIR, NIR and FT-Raman spectroscopic techniques referring to a lactic acid fermentation shows the best prediction performance for FTIR [22].

Some studies described the analysis of glucose, acetate, ammonium and phosphate concentrations in bioprocesses, using ATR-FTIR spectroscopy. Among other substances, ammonium and glucose are analysed in a complex antibiotic fermentation by at-line measurements on a horizontal attenuated total reflectance (HATR) crystal using a partial least squares (PLS) calibration model [23]. Gluconacetobacter xylinus fed-batch cultures were monitored by an in situ ATR probe aimed at the online PLS analysis of acetate, phosphate and ammonium [24]. The results of these PLS predictions—in contrast to MCR-ALS predictions of glucose, acetate, ammonium and phosphate—are discussed below.

As shown by references, FTIR monitoring of bioprocesses and MCR analysis of complex mixtures such as fermentation broths promise many advantages in simultaneous process information collection. The proposed effective usage of in situ and ex situ online sensor data to calculate carbon mass balance-constrained MCR predictions of several analytes underlines the relevance of ATR-FTIR/MCR-ALS combinations in fermentation analysis.

The monitored substances are related to bacterial metabolism. Glucose, ammonia and phosphate are substrates whereas acetate is a by-product of overflow metabolism [25]. The recombinant cytochrome p450 remains unconsidered for its being an intracellular metabolite and therefore not being obtainable in fermentation broth. To predict concentrations of the observed analytes, only four calibration measurements of pure components are required, provided adequate constraints are applied during the alternating least squares procedure. In addition to the required analyte spectra, the implementation of estimated artefact spectra and additionally known fermentation media components is useful. As mentioned above, collecting samples from fermentation broth by a peristaltic pump through a tube system can present the problem of air bubbles in ATR flow-cell with impact on the measured spectra. If the water spectrum is removed from each mixture spectrum prior to that, the air bubble disturbance has its own spectral signature and can be handled like any pure component in multivariate curve resolution. The shape of this artefact is easy to be determined and its implementation in MCR improves the resolution of primary signals, as shown below. All initial estimations for pure spectral components **S** expected in the mixture are also implemented as soft-constraints during alternating least squares. In the following, soft-constraints means the presence of an allowed solution area in a range set by inequality constraints, as in the optimisation problem. In the case of physical spectral shifting in the mixture, a certain flexibility during iterative identification of pure components helps to avoid over-restriction.

In the concentration estimation step, besides the non-negativity constraint, online process data such as input and exhaust gas-flow, fermenter mass, liquid supply from feed reservoir and pH control as well as turbidity are utilized to calculate elementary carbon mass balance constraints. Mass balance constraints (closure) applied to reaction systems have been described [1,26,27]. Closure constraints require invariance of total concentration, granted by the sterile online sampling system and by including total reactor in- and output mass flow in the constraint calculation. So far, as known, the presented application of the carbon mass balance constraint for MCR to analyse a fed-batch fermentation process is a new utilisation of the popular closure constraint. The carbon mass balance constraint for a fed-batch fermentation requires extensive prior calculations, such as different conversion steps and soft sensor approaches. In addition, dynamic in- and output carbon mass-flow in gas and liquid phase as well as continuous and discrete sampling need to be taken into account. The presented algorithm is able to deal with these requirements. Calculating carbon mass balances or recovery rates is an established approach in bioprocess engineering to check the integrity of observed process data [28]. The referenced literature has already shown the application of carbon mass balances for *Escherichia coli* high-cell-density fed-batch culture and recombinant protein production. During process runtime, carbon recovery rates should take on values of about 1. In the present study, this condition is utilised as a MCR-ALS constraint for the estimation of carbon sources and metabolites such as glucose and acetate in fermentation media. In so doing, the explorative decomposition of measured mixture spectra is coupled with analytical knowledge in order to form a new hybrid multivariate modelling approach.

In summary, the objectives of this paper are

(1) the interpretation of the MCR-ALS closure constraint as a carbon mass balance constraint for fed-batch fermentation processes;
(2) to demonstrate that moderate gas bubble disturbances on the ATR crystal can be handled computationally, without any need for technical preventions;
(3) to show that MCR-ALS with carbon mass constraint is capable of simultaneously predicting four analyte concentrations from FTIR spectra of fermentation media samples, with minor calibration effort.

2. Material and Methods

2.1. Spectra Acquisition, Sampling and Spectra Processing

The MIR spectra are scanned with a Thermo Scientific Nicolet™ iS™ 10 and the extension unit Nicolet iZ™ 10. The Specac's Gateway™ ATR Accessory Kit and a ZnSe ATR crystal with six reflections are mounded as a flow cell in that unit. The flow cell is connected to the bioreactor via PTFE tubes (id 1.1 mm) and a Flownamics® FISP® probe with rapid flow membrane for cell-free sampling. A peristaltic pump, controlled by an Arduino microcontroller and a driver board, delivers the sample liquid continuously to the FTIR flow cell. A background spectrum with pure water in the flow cell is scanned before using the FTIR for bioprocess analysis. During a running sampling process, spectra are scanned in cycles of 10 min. During spectrum acquisition by the Thermo Scientific™ OMNIC™ software [29], the sampling pump remains inactive. The spectra acquisition time for scanning 32 spectra and releasing the mean spectrum for the current sample is about 1 min. OMNIC and microcontroller are both triggered by a C# program that observes and synchronises the sample supply and measurement steps. Before each start of a fermentation trial, tubing and flow cell are treated with 70% ethanol solution to minimise the risk of microbial activity in the sampling section. The initial spectra of known substances are standardised to unit concentration. No further pre-processing steps such as normalisation or differentiation are applied to the mixture-spectra in order to preserve the natural physical properties of the spectra. After the fermentation run, the MCR-ALS analysis of the ex situ online-monitored FTIR spectra is performed for all collected spectra.

2.2. Reference Analysis

To validate MCR-ALS results, reference values for glucose, acetate, ammonia and total phosphate concentrations are measured in the cell-free sample drain after passing the FTIR flow-cell. Glucose analysis was conducted by the YSI 2700 SELECT Biochemistry Analyzer (Yellow Springs, OH, USA). Acetate was determined by HPLC (high performance liquid chromatography) using chromatography column Reprogel H+ (Dr. Maisch GmbH, Ammerbuch, Germany). Total phosphate and ammonia were determined using photometric methods by procedures described in DIN EN 1189, DVGW W 504 and DIN 38406 E5.

2.3. Bioreactor System and Online Measurement Equipment

Fermentations are conducted in a prototype of Bioengineering's 5l rounded-bottom autoclavable laboratory fermenter (RALF), controlled and observed with the Software BioSCADA Lab (Bioengineering AG, Wald, Switzerland). A supply tower with intelligent front modules (IFM) directs in- and output of control and measurement values. All data interchanged between IFMs and SCADA pass a structured query language (SQL) data base, the central data hub. From there, the needed data for calculating MCR constraints or advanced measurement and control strategies can be acquired by MySQL and MATLAB.

The current work utilises the following bioprocess online measurement instrumentation: Turbidity probe ASD19-N and optek-converter FC10 (optek-Danulat GmbH, Essen, Germany); exhaust gas analyser BlueInOne Ferm (BlueSens GmbH, Herten, Germany); thermal mass flow controller Red-Y Smart for inflow oxygen (0.01, ... , 5 lpm) and air/nitrogen (0.1, ... , 10 lpm) control (Vögtlin Instruments AG, Aesch, Switzerland); balances for online weight/volume observation of fermenter (DE 35K5D, Kern & Sohn GmbH, Balingen,Germany), acid/base (EW6000-1M, Kern & Sohn GmbH, Balingen, Germany) and feed (BL6100, Sartorius, Göttingen, Germany) reservoir.

2.4. Fermentation Strategy

The HCDC process is conducted in three phases: an initial batch phase, a feeding phase for biomass growth and an induction phase for product expression. The substrate and inductor feed is performed by exponential feeding strategy to control the cell specific growth rates μ similar to [30]. Because of the risk of overflow metabolism und protein folding errors at high growth rates, μ is controlled to defensive setpoints of 0.1 h^{-1} (feed phase/biomass production) and 0.05 h^{-1} (induction phase).

2.5. Strain and Fermentation Medium

E. coli BL21 (DE3) pET28a was stored as glycerol cryo-culture at -76 °C. The pre-culture is incubated as overnight culture in 500 mL baffled flasks at 37 °C in a shaker rotating 200 rpm. An amount of 300 mL pre-culture is portioned in equal shares on two shaking flasks. After 24 h pre-culture incubation, 2.7 L sterilised batch medium in the reactor is inoculated with the culture, thus amounting to a total start volume of 3 L.

The media are modified mineral media based on [31]. The pre-culture and batch medium contain per litre: Glucose*H_2O, 16.5 g; KH_2PO_4, 13.3 g; $(NH_4)_2HPO_4$, 4 g; citric acid, 1.7 g; $MgSO_4$*$7H_2O$, 0.72 g; Fe(II)SO_4*7 H_2O, 113.5 mg; $CoCl_2$*$6H_2O$, 10.5 mg; $MnCl_2$*4 H_2O, 15 mg; $CuCl_2$, 1.2 mg; H_3BO_3, 3 mg; Na2MoO4*2 H2O, 2.5 mg; thiamine*HCl, 4.5 mg; trisodium citrate dihydrate, 75 mg; Na_2-EDTA, 9.6 mg.

The feeding solution is composed of Glucose*H_2O, 544.4 g; $MgSO_4$*$7H_2O$, 12 g; Fe(II)SO_4*$7H_2O$, 43.3 mg; $CoCl_2$*$6H_2O$, 21.4 mg; $MnCl_2$*$4H_2O$, 23.5 mg; $CuCl_2$ 2.5 mg; H_3BO_3, 5 mg; Na_2MoO_4*$2H_2O$, 4 mg; trisodium citrate dihydrate, 116 mg; Na_2-EDTA, 14.8 mg.

3. Theory and Calculation

3.1. Nomenclature

Matrices: Uppercase fat letters
Vectors: Lowercase fat letters
Scalars: Lowercase letters

3.2. Multivariate Curve Resolution and Its Physical Interpretation

The bilinear model of multivariate curve resolution [1] for FTIR data can be deduced from the Lambert–Beer law which describes the attenuation of light travelling through material. The absorbance x of a material is given as

$$x = \lg\left(\frac{I_0}{I_1}\right) = c\,\varepsilon\,d$$

The logarithm of incident radiant intensity (I_0) divided by transmitted radiant intensity (I_1) is equal to the product of substance concentration (c), the molar attenuation coefficient (ε) and the pathlength (d). In this work, the technique of attenuated total reflection is used, so d is the penetration depth of an evanescent wave into the sample on the ATR crystal. The material and wavelength dependent factors ε and d can be pooled to s which consolidates the optical properties of a substance:

$$x = c\,s$$

For mixtures of several substances $k = 1, \dots, \Omega$, each absorbance value x_{ij} related to its wavelength in a spectrum $j = 1, \dots, n$ for a particular concentration profile $i = 1, \dots, m$ is calculated as

$$x_{ij} = \sum_{k=1}^{\Omega} c_{ik}\,\varepsilon_{kj}d_{kj} = \sum_{k=1}^{\Omega} c_{ik}s_{kj}$$

In chemometrics, it is usual to term $i = 1, \dots, m$ as the objects or samples of a dataset, whereby j counts the n features or variables. Here, the m objects are samples of fermentation broth over process runtime and the n features are absorbance values over the wavenumbers of FTIR spectra.

According to the previous sum equation, the decomposition of absorbance values over sample and wavenumber can be organised in matrices

$$\begin{pmatrix} x_{11} & x_{12} & \cdots & x_{1n} \\ x_{21} & x_{22} & \cdots & x_{2n} \\ \vdots & \vdots & \ddots & \vdots \\ x_{m1} & x_{m2} & \cdots & x_{mn} \end{pmatrix} = \begin{pmatrix} c_{11} & c_{12} & \cdots & c_{1\Omega} \\ c_{21} & c_{22} & \cdots & c_{2\Omega} \\ \vdots & \vdots & \ddots & \vdots \\ c_{m1} & c_{m2} & \cdots & c_{m\Omega} \end{pmatrix} \begin{pmatrix} s_{11} & s_{12} & \cdots & s_{1n} \\ s_{21} & s_{22} & \cdots & s_{2n} \\ \vdots & \vdots & \ddots & \vdots \\ s_{\Omega 1} & s_{\Omega 2} & \cdots & s_{\Omega n} \end{pmatrix}$$

In matrix representation, we get the simplified description:

$$\mathbf{X} = \mathbf{C}\mathbf{S}^{T}$$

That is the decomposition of absorbance spectra indicated by multivariate curve resolution assuming the data matrix $\mathbf{X}$ is bilinear.

3.3. An Implementation of the Alternating Least Squares Algorithm

With an initial estimation for concentration matrix $\hat{\mathbf{C}}_0$ or pure components $\hat{\mathbf{S}}_0$ and existing data $\mathbf{X}$, the ALS algorithm can run and perform multivariate curve resolution iteratively [1]. Assuming the chemical rank of the observed data matrix is estimated and one assumption per each expected spectral independent component is available, the ALS procedure can start with an initial pure component matrix. Thus, in the first iteration, the estimated unconstrained concentration matrix $\hat{\mathbf{C}}$ is obtained by

$$\hat{\mathbf{C}} = \mathbf{X}\hat{\mathbf{S}}_0 \left(\hat{\mathbf{S}}_0^{T}\hat{\mathbf{S}}_0 \right)^{-1} = \mathbf{X}\left(\hat{\mathbf{S}}_0^{T} \right)^{+}$$

whereby the superscripted + indicates the pseudoinverse.

In the next step, $\hat{\mathbf{S}}^{T}$ is estimated in an unconstrained way by

$$\hat{\mathbf{S}}^{T} = \left(\hat{\mathbf{C}}^{T}\hat{\mathbf{C}} \right)^{-1} \hat{\mathbf{C}}^{T}\mathbf{X} = \hat{\mathbf{C}}^{+}\mathbf{X}$$

With that pure component estimation, a new concentration matrix calculation can be performed. That loop is repeated until a termination criterion is achieved.

Because of rotational and intensity ambiguities, it is necessary to constrain the solutions for $\hat{\mathbf{C}}$ and $\hat{\mathbf{S}}^{T}$ to obtain a physically meaningful separation of mixture components.

Bioengineering **2017**, *4*, 9

To calculate constrained linear least-squares solutions in this work, the lsqlin function with the *active-set* algorithm from MATLAB and the "Optimization Toolbox" is applied [32]. lsqlin makes use of mathematically rigorous methods of applying equality and inequality constraints with a better numerical stability than approximate methods commonly used in chemometrics. The approximate methods are easy to use and code, but they exhibit poor least squares behaviours and in some cases they result in an increase in the magnitude of residuals [33].

lsqlin solves linear least-squares curve fitting problems of the form

$$\min_{c^T} \| \, x^T - S \, c^T \, \|_2^2 \text{ such that } \begin{cases} A c^T \leq b \\ A_{eq} c^T = b_{eq} \\ l \leq c^T \leq u \end{cases}$$

Hence, the MCR-ALS algorithm using lsqlin is implemented as shown in Figure 1 to solve the present problem of resolving X in a hybrid modelling way with the target of reducing ambiguities of least squares solutions. To bring a priori knowledge about pure spectra and the bioprocess into ALS solutions, linear inequality constraint vectors (e.g., b) and matrices (e.g., A) are applied. Further, the non-negativity constraint for concentrations is set by using lower bounds (l).

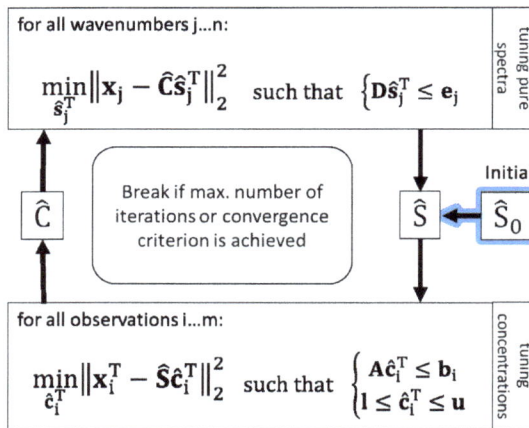

Figure 1. Scheme of the MCR-ALS algorithm with inequality constraints applying MATLAB lsqlin function.

3.4. Constraints for Pure Spectral Component Estimation

Pure spectra of components which are known and expected in mixture are constrained based on measured und normalised spectra of respective pure substances. Therefore, the same spectra used as initial estimations $\hat{S}_0$ are also basis values of inequality constraints to calculate $\hat{S}$. Assuming the shapes of pure spectra in mixture closely resemble the pure measured spectra, in each iteration the associated pure component estimations may only vary inside the defined ranges relative to the measured $\hat{S}_0$. Depending on the amount of expected deviation in mixture, the range for the respective component can be adapted. In so doing, over-restriction can be avoided e.g., in the case of smaller rates of band shifting or in the case of differences in signal-to-noise ratios between high concentrated pure substance measurement and lower concentrations in mixture.

Regarding the inequation constraint for tuning $\hat{S}$ in Figure 1 (upper box), D is composed vertically of the positive and negative identity matrices I and $-I$ both of dimension (Ω, Ω).

$$D = \begin{pmatrix} +I \\ -I \end{pmatrix}$$

The positive part is associated with the upper bounds e_j^u, the negative with the lower bounds e_j^l represented in e_j for all components on each wavenumber. The allowed upper deviation u and lower deviation l are relative to the total ranges of the minimal and maximal values of the pure initial spectra for each component $(\hat{s}_0)_{k=1,...,\Omega}$.

$$\Delta_s = \begin{pmatrix} \max((\hat{s}_0)_{k=1}) \\ \vdots \\ \max((\hat{s}_0)_{k=\Omega}) \end{pmatrix} - \begin{pmatrix} \min((\hat{s}_0)_{k=1}) \\ \vdots \\ \min((\hat{s}_0)_{k=\Omega}) \end{pmatrix}$$

$$e_j = \begin{pmatrix} +e_j^u \\ -e_j^l \end{pmatrix} = \begin{pmatrix} (\hat{s}_0)_j^T + u \circ \Delta_s \\ (\hat{s}_0)_j^T - l \circ \Delta_s \end{pmatrix}$$

In our application, the chemical rank of the mixtures X was estimated at 12 significant spectroscopically independent components by principal component analysis (PCA). The loadings of PCA were manually evaluated for the presence of spectra-like structure, which is strongly present on the first principal components and decreases on higher factors. Among the above mentioned, significant spectroscopically independent components were spectra of known media components, expected metabolic products, artefacts (like air bubbles) and unknown components. Only components evaluated as certainly present in the spectral mixture X are constrained, notably the pure spectra of wanted substances: glucose, ammonia, total phosphate ($H_2PO_4^-$ + HPO_4^{2-}) and acetate. The estimations for those pure components may take on values at an interval of $\pm 10\%$ in the range of each pure component spectrum starting on the initial spectrum (see Table 1).

Table 1. Columns in $\hat{S}_0$ with initial spectra and constraint settings.

	$(\hat{s}_0)_1$	$(\hat{s}_0)_2$	$(\hat{s}_0)_3$	$(\hat{s}_0)_4$	$(\hat{s}_0)_5$	$(\hat{s}_0)_6$	$(\hat{s}_0)_7$	$(\hat{s}_0)_8$	$(\hat{s}_0)_9$	$(\hat{s}_0)_{10}$	$(\hat{s}_0)_{11}$	$(\hat{s}_0)_{12}$
pure	glc	ace	NH_4^+	$H_2PO_4^-$ HPO_4^{2-}	IPTG	$MgSO_4$	air	H_2O	citric acid	unifrnd	unifrnd	unifrnd
u	0.1	0.1	0.1	0.1	0.1	0.1	inf	inf	0.1	inf	inf	inf
l	−0.1	−0.1	−0.1	−0.1	−0.1	−0.1	−inf	−inf	−0.1	−inf	−inf	−inf

Initial pure components: glucose (glc); acetate (ace); isopropyl β-D-1-thiogalactopyranoside (IPTG); air bubbles estimation (air); uniformly distributed random values (unifrnd); infinity (inf).

Because of water background subtraction on each taken spectrum x_i, air bubbles in the flow cell have the shape of inverted water spectra. Moreover, a pure water spectrum is also initialised for the case of air bubble presence during the background recording. Both estimations are not constrained and can vary depending on the actual mixture content.

To demonstrate the validity of the assumption for the spectral air bubble model, a simple aqueous solution containing glucose ($15 \cdot gL^{-1}$), ammonium ($0.7\ gL^{-1}$) and phosphate ($8 \cdot gL^{-1}$) was compounded. From this solution, a first FTIR spectrum was acquired from a mixture covering the entire ATR crystal whereas a second spectrum resulted from the same mixture covering only about half of the crystal surface. In this way, a part of the IR beam reflections interacts with the aqueous solution on the ATR crystal, while another part interacts just with the air on the crystal surface. The latter liquid-free surface part simulates a large air bubble on the crystal in ATR flow cell. In the case without ALS iterations, the mixture matrix of the known solution is multiplied once with a simple pseudo inverse of the estimated initial pure components matrix. In a first measurement, S_0 contains just pure measurements of glucose, ammonium and phosphate, respectively standardised to unit concentration. Next, S_0 additionally contains an inverted water spectrum. The differences in concentration estimations are shown in Figure 2 whereas the actual concentration values are listed in

Table 2. Obviously, the integration of the air bubble model brings an improvement of the prediction results in the case of air bubble presence.

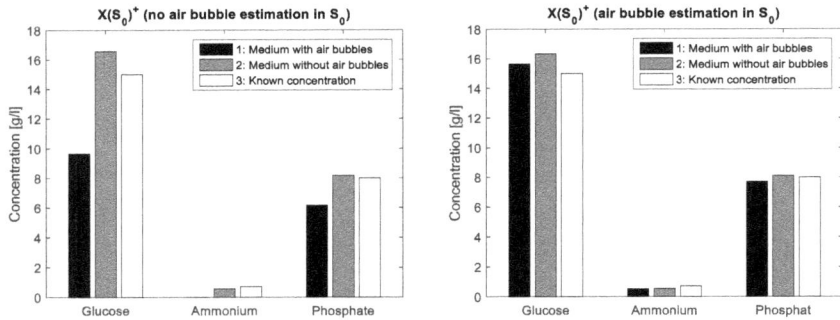

Figure 2. Proof of the initial estimation for the spectral air bubbles signature: in a known aqueous solution of glucose, ammonium and phosphate, air bubbles are present in one case and absent in another. On the left, the known spectral components S_0 do not consider the air bubble model. On the right, the air bubble model is integrated and the differences between concentrations estimated in medium with and without air bubbles are similarly close to the known concentrations.

Table 2. Concentration estimation by multiplying known mixture spectra with pseudo inverses of pure spectral component matrices with and without integration of an air bubble model.

	Air Bubble Model in S_0	Glucose $[\text{gL}^{-1}]$	Ammonium $[\text{gL}^{-1}]$	Phosphate $[\text{gL}^{-1}]$
Medium with air bubbles	No	9.65	0	6.17
	Yes	15.63	0.52	7.70
Medium without air bubbles	No	16.56	0.57	8.16
	Yes	16.32	0.55	8.10
Known concentration	-	15	0.7	8

3.5. Constraints for Concentration Estimation

In bioprocess engineering, the carbon balance and recovery rate of a fermentation process are commonly used as a check for the integrity of process monitoring and sensors as well as the assessment of the release of outer membrane components. Carbon balances in a fed-batch culture are based on the mass of carbon in the total fermenter volume. Thereby, the recovery rate is the relation between the recovered carbon $m^{C,rec}(t)$ and the carbon brought into the bioreactor $m^{C,in}(t)$ over process runtime t.

$$r^C(t) = \frac{m^{C,rec}(t)}{m^{C,in}(t)}$$

Suppose all carbon compounds are determinable and measurement errors are negligible, $r^C(t)$ is equal to 1 for all t. Because of the presence of measurement errors and not identified soluble organic carbon compounds, a tolerance range must be assumed. The carbon recovery considering biomass, CO_2, glucose and acetate is assumed as being about 90% [28].

The recovered carbon is the sum of carbon mass in the reactor liquid phase L, gas phase g, sample liquid phase divided in cell-free sampling scf and cell containing sampling scc.

$$m^{C,rec}(t) = m^{C,L}(t) + m^{C,g}(t) + m^{C,scc}(t) + m^{C,scf}(t)$$

The brought-in carbon is the sum of initial carbon mass in fermentation medium at process start time (t = 0) and the supplied carbon mass m^r from the feed reservoir r.

$$m^{C,in}(t) = m^{C,L}(t = 0) + m^{C,r}(t)$$

If the integrity of measurement equipment and data observation is already proved, the carbon mass balance can be applied as a MCR-ALS constraint for glucose and acetate estimation from the spectra on each observation i over process runtime. For that, several non-spectroscopic measurements and assumptions must be applied to calculate carbon balances on each FTIR measurement.

The carbon in the reactor liquid phase is located in biomass in fractions of $\alpha^{C,cell}$ as well as in dissolved CO_2 in fractions of α^{C,CO_2}. The fraction of carbon in biomass is an assumption based on the analysis of elemental biomass composition of *E. coli* with an elemental analyser taken from literature [34]. Further carbon, of course, is located in glucose (glc) and acetate (ace), for which the concentrations c in reactor volume V^L are to be determined by FTIR/MCR-ALS. For MCR execution, $m_i^{C,L}$ must be split into one term containing the required concentrations from FTIR ex situ online measurement (ex) and into another term containing carbon compounds with concentrations accessible by in situ online measurement (on). For calculation of the online term, biomass concentration is observed by turbidity measurement and a calibrated exponential model. Dissolved carbon dioxide concentration is estimated by a soft-sensor based on Henry's law and the CO_2 mole fraction measured by a gas sensor in exhaust gas flow [35].

$$m_i^{C,L} = m_i^{C,L,on} + m_i^{C,L,at}$$

$$m_i^{C,L,on} = \left(\alpha^{C,cell} c_i^{cell,L} + \alpha^{C,CO_2} c_i^{CO_2,L} \right) V_i^L$$

$$m_i^{C,L,ex} = \left(\alpha^{C,glc} c_i^{glc,L} + \alpha^{C,ace} c_i^{ace,L} \right) V_i^L$$

The ex situ online term ex, containing the concentrations to estimate by MCR, must be converted concerning the left side of inequality constraint $\mathbf{A}c_i^T \leq \mathbf{b}_i$. Thus, the $(2, \Omega)$-matrix $\mathbf{A}$ contains in the first row, on positions associated with concentrations of glucose and acetate, the fractions $\alpha^{C,glc}$ and $\alpha^{C,ac}$ multiplied with reactor volumes V_i^L. The first row is associated with the upper bounds $\mathbf{b}^u$ of the constraint. The second row is the negative of the first row and is associated with the lower bounds $\mathbf{b}^l$.

All other brought-in and recovery terms that are directly or indirectly accessible by online process sensors and soft-sensors, but not by FTIR/MCR-ALS, are used to form the $\mathbf{b}_i$ vector.

The carbon in the exhaust gas phase is calculated by the CO_2 removal rate Q^{CO_2}, which in turn is calculated based on measurements of CO_2 mass flow at gas phase entry and of inert gas balance to estimate exit mass flow.

$$m_i^{C,g} = \alpha^{C,CO_2} \int_1^i V_i^L(t) Q_i^{CO_2}(t) dt$$

The calculation of carbon in samples of fermentation media starts at the first FTIR observation i = 1 with known initial media concentrations c_0 and sample volumes ΔV_0^{scc} and ΔV_0^{scf} taken before the first FTIR measurement is observed. A certain error in sample carbon mass calculation must be accepted since the respective current values of $c_i^{glc,L}$ and $c_i^{ace,L}$ are unknown at the time of constraint calculation. Hence, at i > 1, the results of the last MCR step i-1 are utilised. Considering comparative slow bioprocess kinetics and a higher sampling frequency, this is a reasonable approximation.

$$m_i^{C,scc} = \sum_{i=1}^m \left(\alpha^{C,glc} c_{i-1}^{glc,L} + \alpha^{C,ace} c_{i-1}^{ace,L} + \alpha^{C,cell} c_{i-1}^{cell,L} + \alpha^{C,CO_2} c_{i-1}^{CO_2,L} \right) \Delta V_{i-1}^{scc}$$

$$m_i^{C,scf} = \sum_{i=1}^m \left(\alpha^{C,glc} c_{i-1}^{glc,L} + \alpha^{C,ace} c_{i-1}^{ace,L} + \alpha^{C,CO_2} c_{i-1}^{CO_2,L} \right) \Delta V_{i-1}^{scf}$$

The brought-in carbon is the sum of the carbon fractions of glucose, acetate and cell mass in the initial medium as well as the supplied glucose from the feed reservoir.

$$m_i^{C,in} = m_{i=1}^{C,L} + m_i^{C,r} = \left(\alpha^{C,glc}\,c_{i=1}^{glc,L} + \alpha^{C,ace}c_{i=1}^{ace,L} + \alpha^{C,cell}c_{i=1}^{cell,L}\right)V_{i=1}^{L} + \alpha^{C,glc}\,c_i^{glc,r}V_i^{r}$$

By that information, $\mathbf{b}_i$ can be calculated as

$$\mathbf{b}_i = \left(\begin{array}{c} b^u \cdot m_i^{C,in} - \left(m_i^{C,L,on} + m_i^{C,g} + m_i^{C,scc} + m_i^{C,scf}\right) \\ -b^l \cdot m_i^{C,in} + \left(m_i^{C,L,on} + m_i^{C,g} + m_i^{C,scc} + m_i^{C,scf}\right) \end{array} \right)$$

The settings of the upper and lower tolerance bounds b^u and b^l of the carbon balance constraint are based on different considerations. Recovery rates higher than 1 are only caused by measurement errors while values below 1 are caused by both, measurement errors and not identified by-products. Therefore, the upper bound can be set tighter than the lower, with values in an interval of $b^u = (0.9, 1.1)$, depending on the process phase. An upper bound lower than 1 may be suitable if it is evident that the carbon compounds which are considered in the constraint calculation but which lie outside the optimisation of $\hat{\mathbf{C}}$ are underestimated (e.g., biomass). An upper bound higher than 1 is indicated if external carbon compounds seem to be overestimated. Thus, by tuning b^u, it therefore is possible to compensate measurement errors in online sensor equipment. As mentioned above, the recovery rate can reach approximately 90% at the end of the *E. coli* process, although just based on glucose, acetate, CO_2 and biomass. In order to take the formation of not considered carbon compounds into account, the lower bound is set in a defensive way to $b^l = 0.8$ to avoid over-restriction.

Furthermore, the non-negativity constraint is set for all concentration values, and at each new curve resolution step, the start value for lsqlin optimisation is set to the last estimation result. The initial concentration values for glucose, acetate, total phosphate and ammonia at $i = 1$ are set to the known batch medium concentration and may vary in a range of $\pm 10\%$.

Some online measurements such as fermenter weight, gas analysis and turbidity have a higher noise level and have to be filtered before further processing. Biomass estimated by turbidity is smoothed by application of an exponential smoothing filter with a smoothing factor alpha set to 0.05. The online signal of fermenter weight is prone to disturbances in form of high needle peaks, often caused by manual contact with the reactor e.g., while taking an offline sample. Those disturbances can easily be removed automatically by a threshold filter detecting differences between one measurement point to the next, higher than a threshold value e.g., >1.5 L, since offline samples usually have values below 1.5 and since the actual fermenter volume changing rate is much inferior. These few values in the sequel above the threshold are overwritten by the last value lower than the threshold. In this way, measurement errors can be significantly reduced since all concentration values are depending on the reactor volume. Outliers in the online gas analysis are treated correspondingly.

4. Results and Discussion

The carbon balance constraint algorithm with appropriate initial pure spectra estimations results in physically reasonable MCR solutions. By setting suitable start values and tolerance bounds, the rotatory and intensity ambiguities are reduced significantly. As a consequence, the concentration profiles of the substrates glucose, ammonia, total phosphate and the expected metabolic by-product acetate can be unfolded from the spectral mixture matrix $\mathbf{X}$ with minor manual measurement effort. An overview of the entire process spectra is displayed in Figure 3.

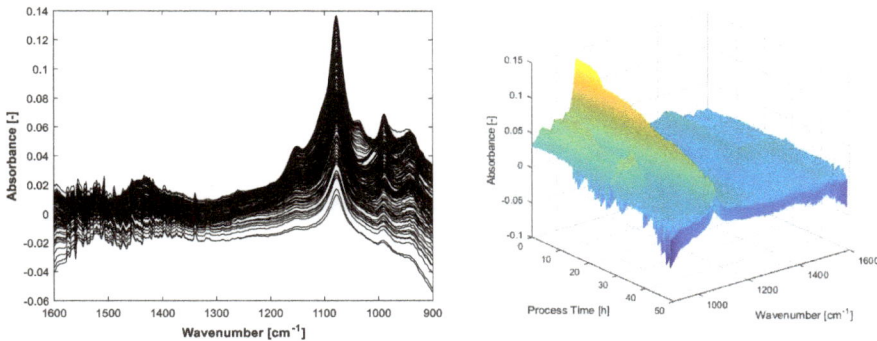

Figure 3. The 264 infrared spectra acquired from approximately 50 h fed-batch fermentation.

The FTIR spectra show negative values because of the water background subtraction. The inflexions downwards on the left and right borders of the display are caused by air bubbles in the flow cell. These artefacts can be handled by MCR. Before integrating the spectral air bubble model in the MCR-ALS algorithm, the assumption for the spectral air bubble model was ascertained by concentration prediction of a simple aqueous solution containing glucose, acetate and phosphate. The solution was measured by FTIR with and without air on the ATR crystal surface. The prediction was executed by multiplying the measured spectra X with the pseudoinverse of S_0, whereby S_0 is a composition of pure spectra of known mixture components. In one experiment, S_0 involves an estimated spectral model for air bubbles, in the other just the pure spectra of the solved components are compounded. As evident from Figure 2 and Table 2, the integration of the estimated air bubble signature results in a significant prediction improvement.

The results of MCR-ALS concentration prediction based on the 264 measured process spectra are shown in Figure 4. Elapsed calculation time for 300 ALS iterations was about 10 min on an Intel Core i7-4790 @3.6 GHz (4 Cores). It should be noted that, besides the constraints described above, just single manual measured spectra for each estimated pure component are utilised to achieve the resolution. Likewise, some of the pure component start values are just vectors of uniformly distributed random numbers. After 300 alternating least squares iterations, satisfactory approximations of the process dynamics are obtained. Glucose and phosphate are present in higher concentrations, so the resolution succeeds nearly without artefacts. At concentrations close to zero, a higher presence of artefacts and noise is expectedly obtained. Accordingly, the lower concentrated ammonium and acetate show a higher ratio of disturbances.

The error evaluation takes place by comparing the FTIR/MCR-ALS concentration measurements with reference measurements. Concerning the residuals, the root mean squared errors (RMSE) are calculated and shown in Table 3.

The prediction results of the proposed MCR-ALS algorithm can be compared with prediction performances of PLS models. Acetate, ammonium and phosphate concentrations of a *Gluconacetobacter xylinus* fed-batch culture were predicted from spectra of in situ ATR-FTIR measurements by a PLS model with accuracies of 0.2, 0.17 and 0.24 gL^{-1}, respectively [24]. The validation errors for offline samples of the same process were 0.22 gL^{-1} (acetate), 0.24 gL^{-1} (ammonium) and 0.18 gL^{-1} (phosphate). The applied PLS regression model is based on 56 mixture solutions, used as calibration standards. The accuracies of MCR-ALS estimation for ammonium and acetate are similar to the PLS errors of the referenced paper. The absolute error of phosphate prediction is higher for the MCR-ALS approach than for the described PLS method, the measurement range being about two times higher, too. In consideration of the minor calibration effort of the proposed MCR-ALS approach, the results are impressive. Furthermore, the PLS glucose prediction accuracy by at-line ATR-FTIR monitoring of an antibiotic fermentation process is with 0.56 gL^{-1} similar to the present prediction by MCR-ALS [23]. The PLS calibration model for glucose is based on 70 filtrated fermentation samples. Here, too, the reduction of calibration effort by effective online sensor data usage is evident when compared to PLS.

The estimations of pure component spectra are displayed in Figure 5. In addition to the notice concerning the associated concentrations, the higher noise level of the lower concentrated ammonium and acetate is also apparent in the pure spectral components.

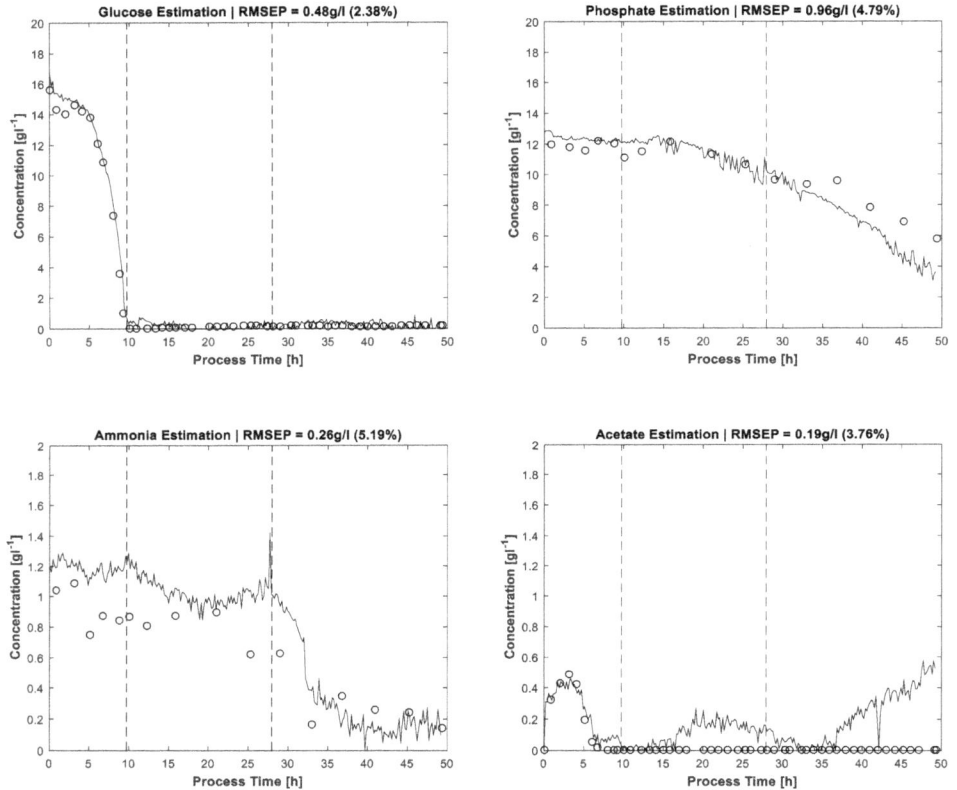

Figure 4. Concentration profiles estimated by FTIR/MCR-ALS (solid lines) and reference measurements (dots). The two dashed vertical lines in each plot distinguish the three process phases: batch phase (left), feeding phase/biomass production (middle), induction phase (right).

Table 3. Prediction performance: error estimation quantified by root mean squared distances between FTIR-MCR-ALS predictions and reference measurements.

Substance	RMSE [gL^{-1}]	rel. RMSE [%]	Expected Range [gL^{-1}]
Glucose	0.48	2.38	(0, 20)
Phosphate	0.96	4.79	(0, 20)
Ammonium	0.26	5.19	(0, 5)
Acetate	0.19	3.76	(0, 5)

Root mean squared error (RMSE); relative RMSE related to expected range (rel. RMSE).

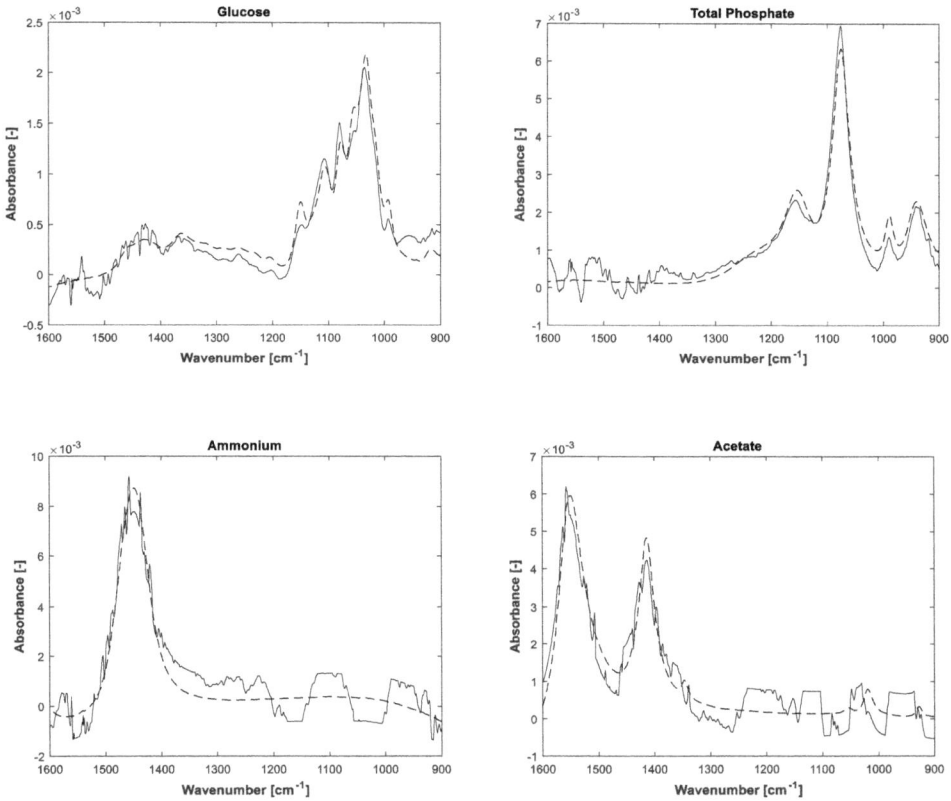

Figure 5. Initial measurements of pure substances (dashed lines) and identified pure spectral components after 300 ALS iterations (solid lines).

By way of comparison, Figure 6 shows the results for glucose and acetate concentrations without application of the carbon balance constraint but including the same constraints as used for pure spectra estimation, see above. Between hour 10 and 15 there is a significant artefact observable in the glucose concentration profile. The concentration estimation is too high, also discernible by carbon recovery rates approaching almost 1.2 in this process phase. A second drift in glucose concentration is located around t = 35 h without obvious reflection in carbon balance because of the lower deviation. In any case, without the carbon balance constraint, the solution space of MCR is enlarged, thereby also increasing the risk of ambiguities which can cause physically nonsensical solutions. For the same reason, the acetate profile in Figure 6 gives the impression of increasing concentrations which actually are not present. Nevertheless, the shapes of the actual existing concentration profiles in the batch phase are more or less recognised, the artefacts increasing mostly in the respective zero-concentration phases.

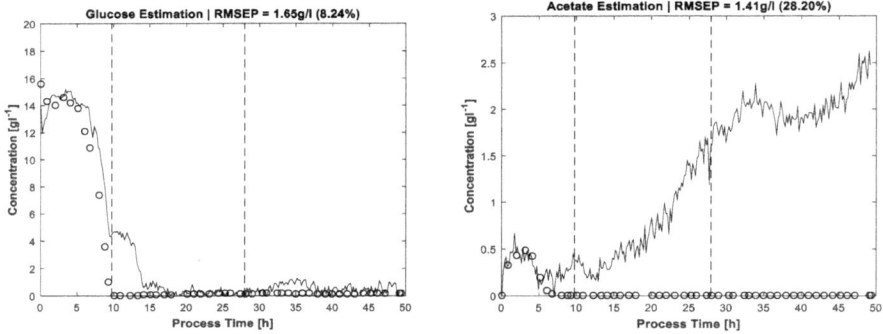

Figure 6. Concentration profiles for glucose (left) and acetate (right), estimated without application of the carbon balance constraint.

The carbon recovery at the end of the MCR-ALS procedure is shown in Figure 7. Without application of the carbon balance constraint, the recovery rate exceeds two times the value of 1.09, once in the beginning and once again at the end of fed-batch phase. Even the lower value of 0.8 is slightly undershot at the beginning of the process. Around t = 25h, near the end of the feeding phase, the recovery rate with the enabled carbon balance constraint touches the highest upper bound of 1.09. As for the final estimated concentrations, the lower bound of 0.8 is not reached.

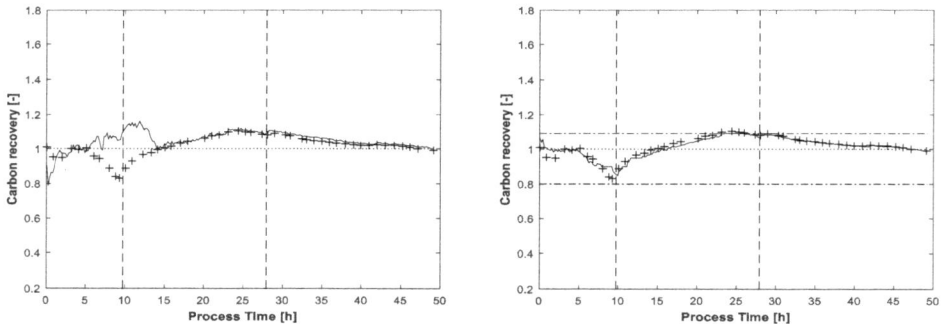

Figure 7. Carbon recovery based on the final glucose and acetate estimations after 300 ALS iterations, without (left) and with (right) application of the carbon balance constraint (solid line). Carbon recovery calculated by offline determined glucose and acetate concentrations (+). The dot-dashed horizontal lines indicate the highest upper and lowest lower bounds, only enabled in the run displayed on the right. The dotted horizontal line is the theoretical recovery rate of 1. The vertical lines indicate the three process phases: batch, feeding and induction.

5. Conclusions

This study has shown that MCR-ALS with tailored constraints is capable of analysing simultaneously the concentrations of glucose, acetate, ammonium and total phosphate from ex situ online recorded FTIR spectra of an *E. coli* HCDC fermentation process. The required concentration information, extracted from 264 FTIR spectra and recorded over 50 h process time, has been estimated in accuracies between 0.19 and 0.96 gL^{-1}. These results are comparable to established concentration estimations by PLS models, but are achieved with less calibration effort. It became apparent that the application of appropriate constraints, in particular the carbon balance constraint, improves the accuracy of concentration estimation in the ALS solution process by avoiding artefacts

Bioengineering **2017**, *4*, 9

caused by rotatory ambiguities. In MCR-ALS concentration estimation, the carbon mass balance constraint, calculated by online sensor data, reduced ambiguities in glucose and acetate concentrations significantly. In pure spectra estimation, initial FTIR measurements of the required analytes as well as a spectral air bubble model led to appropriate MCR solutions. Besides the automatically sampled online FTIR spectra, all applied constraints are calculated broadly based on automated measurements and analytical process knowledge. It is shown that by introducing prior knowledge and processed non-spectroscopic online sensor data into the ALS procedure, better spectra resolution performances as well as efficient fermentation process analysis can be achieved.

Acknowledgments: This study was kindly financed within the BMBF Project MAGNENZ, Magnetic Enzymes (FKZ: 0316057B).

Author Contributions: Dennis Vier, Klaus-Uwe Gollmer and Volker Schünemann conceived and designed the experiments; Dennis Vier and Stefan Wambach performed the experiments; Dennis Vier analyzed the data; the University of Applied Sciences Trier–Umwelt Campus Birkenfeld (Departments "Bioprocess Engineering" and "Applied Informatics") contributed reagents/materials/analysis tools; Dennis Vier wrote the paper.

Conflicts of Interest: The authors declare no conflict of interest.

References

1. Tauler, R.; Kowalski, B.; Fleming, S. Multivariate curve resolution applied to spectral data from multiple runs of an industrial process. *Anal. Chem.* **1993**, *65*, 2040–2047. [CrossRef]
2. Zhang, X.; Tauler, R. Measuring and comparing the resolution performance and the extent of rotation ambiguities of some bilinear modeling methods. *Chemom. Intell. Lab. Syst.* **2015**, *147*, 47–57. [CrossRef]
3. Ruckebusch, C.; Blanchet, L. Multivariate curve resolution: A review of advanced and tailored applications and challenges. *Anal. Chim. Acta* **2013**, *765*, 28–36. [CrossRef] [PubMed]
4. De Juan, A.; Jaumot, J.; Tauler, R. Multivariate Curve Resolution (MCR). Solving the mixture analysis problem. *Anal. Methods* **2014**, *6*, 4964–4967. [CrossRef]
5. Abdollahi, H.; Tauler, R. Uniqueness and rotation ambiguities in Multivariate Curve Resolution methods. *Chemom. Intell. Lab. Syst.* **2011**, *108*, 100–111. [CrossRef]
6. Pomerantsev, A.L.; Rodionova, O.Y. Process analytical technology. A critical view of the chemometricians. *J. Chemom.* **2012**, *26*, 299–310. [CrossRef]
7. Azzouz, T.; Tauler, R. Application of multivariate curve resolution alternating least squares (MCR-ALS) to the quantitative analysis of pharmaceutical and agricultural samples. *Talanta* **2008**, *74*, 1201–1210. [CrossRef] [PubMed]
8. Richards, S.E.; Becker, E.; Tauler, R.; Walmsley, A.D. A novel approach to the quantification of industrial mixtures from the Vinyl Acetate Monomer (VAM) process using Near Infrared spectroscopic data and a Quantitative Self Modeling Curve Resolution (SMCR) methodology. *Chemom. Intell. Lab. Syst.* **2008**, *94*, 9–18. [CrossRef]
9. Conti, P.; Zamponi, S.; Giorgetti, M.; Berrettoni, M.; Smyrl, W.H. Multivariate curve resolution analysis for interpretation of dynamic Cu K-edge X-ray absorption spectroscopy spectra for a Cu doped V(2)O(5) lithium battery. *Anal. Chem.* **2010**, *82*, 3629–3635. [CrossRef] [PubMed]
10. Garrido, M.; Rius, F.X.; Larrechi, M.S. Multivariate curve resolution-alternating least squares (MCR-ALS) applied to spectroscopic data from monitoring chemical reactions processes. *Anal. Bioanal. Chem.* **2008**, *390*, 2059–2066. [CrossRef] [PubMed]
11. Jaumot, J.; Vives, M.; Gargallo, R. Application of multivariate resolution methods to the study of biochemical and biophysical processes. *Anal. Biochem.* **2004**, *327*, 1–13. [CrossRef] [PubMed]
12. Piqueras, S.; Duponchel, L.; Offroy, M.; Jamme, F.; Tauler, R.; de Juan, A. Chemometric strategies to unmix information and increase the spatial description of hyperspectral images: A single-cell case study. *Anal. Chem.* **2013**, *85*, 6303–6311. [CrossRef] [PubMed]
13. Blanco, M.; Peinado, A.C.; Mas, J. Monitoring alcoholic fermentation by joint use of soft and hard modelling methods. *Anal. Chim. Acta* **2006**, *556*, 364–373. [CrossRef]

14. Grassi, S.; Amigo, J.M.; Lyndgaard, C.B.; Foschino, R.; Casiraghi, E. Assessment of the sugars and ethanol development in beer fermentation with FT-IR and multivariate curve resolution models. *Food Res. Int.* **2014**, *62*, 602–608. [CrossRef]

15. Grassi, S.; Alamprese, C.; Bono, V.; Casiraghi, E.; Amigo, J.M. Modelling Milk Lactic Acid Fermentation Using Multivariate Curve Resolution-Alternating Least Squares (MCR-ALS). *Food Bioprocess Technol.* **2014**, *7*, 1819–1829. [CrossRef]

16. Koch, C.; Posch, A.E.; Goicoechea, H.C.; Herwig, C.; Lendl, B. Multi-analyte quantification in bioprocesses by Fourier-transform-infrared spectroscopy by partial least squares regression and multivariate curve resolution. *Anal. Chim. Acta* **2014**, *807*, 103–110. [CrossRef] [PubMed]

17. Mazarevica, G.; Diewok, J.; Baena, J.R.; Rosenberg, E.; Lendl, B. On-line fermentation monitoring by mid-infrared spectroscopy. *Appl. Spectrosc.* **2004**, *58*, 804–810. [CrossRef] [PubMed]

18. Landgrebe, D.; Haake, C.; Höpfner, T.; Beutel, S.; Hitzmann, B.; Scheper, T.; Rhiel, M.; Reardon, K.F. On-line infrared spectroscopy for bioprocess monitoring. *Appl. Microbiol. Biotechnol.* **2010**, *88*, 11–22. [CrossRef] [PubMed]

19. Lourenço, N.D.; Lopes, J.A.; Almeida, C.F.; Sarraguça, M.C.; Pinheiro, H.M. Bioreactor monitoring with spectroscopy and chemometrics: A review. *Anal. Bioanal. Chem.* **2012**, *404*, 1211–1237. [CrossRef] [PubMed]

20. Schmitt, J.; Flemming, H.-C. FTIR-spectroscopy in microbial and material analysis. *Int. Biodeterior. Biodegrad.* **1998**, *41*, 1–11. [CrossRef]

21. Shih, C.-J.; Smith, E.A. Determination of glucose and ethanol after enzymatic hydrolysis and fermentation of biomass using Raman spectroscopy. *Anal. Chim. Acta* **2009**, *653*, 200–206. [CrossRef] [PubMed]

22. Sivakesava, S.; Irudayaraj, J.; Ali, D. Simultaneous determination of multiple components in lactic acid fermentation using FT-MIR, NIR, and FT-Raman spectroscopic techniques. *Process Biochem.* **2001**, *37*, 371–378. [CrossRef]

23. Roychoudhury, P.; Harvey, L.M.; McNeil, B. At-line monitoring of ammonium, glucose, methyl oleate and biomass in a complex antibiotic fermentation process using attenuated total reflectance-mid-infrared (ATR-MIR) spectroscopy. *Anal. Chim. Acta* **2006**, *561*, 218–224. [CrossRef]

24. Kornmann, H.; Valentinotti, S.; Duboc, P.; Marison, I.; von Stockar, U. Monitoring and control of Gluconacetobacter xylinus fed-batch cultures using in situ mid-IR spectroscopy. *J. Biotechnol.* **2004**, *113*, 231–245. [CrossRef] [PubMed]

25. Xu, B.; Jahic, M.; Enfors, S.-O. Modeling of Overflow Metabolism in Batch and Fed-Batch Cultures of *Escherichia coli*. *Biotechnol. Prog.* **1999**, *15*, 81–90. [CrossRef] [PubMed]

26. Tauler, R.; Smilde, A.; Kowalski, B. Selectivity, local rank, three-way data analysis and ambiguity in multivariate curve resolution. *J. Chemom.* **1995**, *9*, 31–58. [CrossRef]

27. Diewok, J.; de Juan, A.; Maeder, M.; Tauler, R.; Lendl, B. Application of a Combination of Hard and Soft Modeling for Equilibrium Systems to the Quantitative Analysis of pH-Modulated Mixture Samples. *Anal. Chem.* **2003**, *75*, 641–647. [CrossRef] [PubMed]

28. Han, L.; Enfors, S.-O.; Häggström, L. *Escherichia coli* high-cell-density culture: Carbon mass balances and release of outer membrane components. *Bioprocess Biosyst. Eng.* **2003**, *25*, 205–212. [CrossRef] [PubMed]

29. Thermo Fisher Scientific Inc. *OMNIC 9*. Version 9.1.24; Thermo Fisher Scientific Inc.: Waltham, MA, USA.

30. Korz, D.J.; Rinas, U.; Hellmuth, K.; Sanders, E.A.; Deckwer, W.-D. Simple fed-batch technique for high cell density cultivation of *Escherichia coli*. *J. Biotechnol.* **1995**, *39*, 59–65. [CrossRef]

31. Riesenberg, D.; Schulz, V.; Knorre, W.A.; Pohl, H.-D.; Korz, D.; Sanders, E.A.; Ross, A.; Deckwer, W.D. High cell density cultivation of *Escherichia coli* at controlled specific growth rate. *J. Biotechnol.* **1991**, *20*, 17–27. [CrossRef]

32. The MathWorks, Inc. *MATLAB and Optimization Toolbox Release*; The MathWorks, Inc.: Natick, MA, USA, 2015.

33. Van Benthem, M.H.; Keenan, M.R.; Haaland, D.M. Application of equality constraints on variables during alternating least squares procedures. *J. Chemom.* **2002**, *16*, 613–622. [CrossRef]

34. Kayser, A. Metabolic flux analysis of *Escherichia coli* in glucose-limited continuous culture. I. Growth-rate-dependent metabolic efficiency at steady state. *Microbiology* **2005**, *151*, 693–706. [CrossRef] [PubMed]

35. Schumpe, A.; Quicker, G.; Deckwer, W.-D. Gas solubilities in microbial culture media. In *Reaction Engineering*; Springer: Berlin/Heidelberg, Germany, 1982; pp. 1–38.

bioengineering

MDPI

Article

Impact of Glycerol as Carbon Source onto Specific Sugar and Inducer Uptake Rates and Inclusion Body Productivity in *E. coli BL21(DE3)*

Julian Kopp [1], Christoph Slouka [1], Sophia Ulonska [2], Julian Kager [2], Jens Fricke [1], Oliver Spadiut [2] and Christoph Herwig [1,2,*]

[1]   Christian Doppler Laboratory for Mechanistic and Physiological Methods for Improved Bioprocesses, Institute of Chemical, Environmental and Biological Engineering, Vienna University of Technology, 1060 Vienna, Austria; julian.kopp@tuwien.ac.at (J.K.); christoph.slouka@tuwien.ac.at (C.S.); fricke_jens@gmx.de (J.F.)
[2]   Research Division Biochemical Engineering, Institute of Chemical, Environmental and Biological Engineering, Vienna University of Technology, 1060 Vienna , Austria; sophia.ulonska@tuwien.ac.at (S.U.); julian.kager@tuwien.ac.at (J.K.); oliver.spadiut@tuwien.ac.at (O.S.)
*   Correspondence: christoph.herwig@tuwien.ac.at; Tel.: +43-(1)-58801-166084

Academic Editor: Mark Blenner
Received: 24 October 2017; Accepted: 19 December 2017; Published: 21 December 2017

Abstract: The Gram-negative bacterium *E. coli* is the host of choice for a multitude of used recombinant proteins. Generally, cultivation is easy, media are cheap, and a high product titer can be obtained. However, harsh induction procedures using isopropyl β-D-1 thiogalactopyranoside as inducer are often referred to cause stress reactions, leading to a phenomenon known as "metabolic" or "product burden". These high expressions of recombinant proteins mainly result in decreased growth rates and cell lysis at elevated induction times. Therefore, approaches tend to use "soft" or "tunable" induction with lactose and reduce the stress level of the production host. The usage of glucose as energy source in combination with lactose as induction reagent causes catabolite repression effects on lactose uptake kinetics and as a consequence reduced product titer. Glycerol—as an alternative carbon source—is already known to have positive impact on product formation when coupled with glucose and lactose in auto-induction systems, and has been referred to show no signs of repression when cultivated with lactose concomitantly. In recent research activities, the impact of different products on the lactose uptake using glucose as carbon source was highlighted, and a mechanistic model for glucose-lactose induction systems showed correlations between specific substrate uptake rate for glucose or glycerol ($q_{s,C}$) and the maximum specific lactose uptake rate ($q_{s,lac,max}$). In this study, we investigated the mechanistic of glycerol uptake when using the inducer lactose. We were able to show that a product-producing strain has significantly higher inducer uptake rates when being compared to a non-producer strain. Additionally, it was shown that glycerol has beneficial effects on viability of cells and on productivity of the recombinant protein compared to glucose.

Keywords: *E. coli*; mixed feed system; glycerol; recombinant proteins; bioprocess engineering

1. Introduction

The Gram-negative bacterium *E. coli* is the expression host of choice for the production of 30% to 40% of recombinant drugs in industry [1,2]. As *E. coli* shows very fast replication rates [3,4] on comparatively cheap media [5], the benefits often outweigh the numerous purification steps [1,6] and the missing glycosylation pattern [1,7,8]. Recombinant protein production in *E. coli* gained more interest again as the demand in single-chain antibody fragments increased, which can be properly expressed in *E. coli* [1,8]. The strain *BL21(DE3)*, created by F. Studier and B. Moffatt back in 1986 [9], is often used in an industrial scale because of very low acetate formation, high replication rates as an effect of the integrated T7-polymerase [9–14], as well as the possibility of protein secretion into the fermentation broth due to a type 2 secretion protein [15–17]. As the lac operon is still one of the most favored promotors in pET-expression systems [3,12,18], it is generally used for insertion of the gene of interest. The repressor protein can only be blocked by allolactose or a structural analogue [19], e.g., the well-known inducer isopropyl β-ᴅ-1 thiogalactopyranoside (IPTG) [3,13]. However, induction with IPTG stresses the cells, as IPTG in higher concentrations is referred to be toxic at elevated induction times [13,18,20]. As tunable protein production is commonly applied in industry nowadays, mixed-feed systems using either IPTG [21] or lactose [13,22,23] as inducer did result in higher product yields when compared to other inducer supplies [24]. Soft induction performed with lactose shows promising results [13,23,25]. As lactose can be metabolized in *E. coli*, it does not stress the cells as much as IPTG [26]. For the production of soluble proteins, induction with lactose usually is preferred [3], but it has also been shown that lactose shows promising results for Inclusion Bodies (IBs) and products located in the periplasm [3,27].

IBs have originally been believed to be waste products by bacteria [28], until it was realized that IBs tend to form as a stress reaction by the cells resulting in a biologically inactive protein [29–31]. Stress reactions of the cells can be caused by high temperatures, pH-shifts, or due to high feeding rates. Higher feeding rates result in higher yields of product [1], which of course is advantageous when combined with the possibility of expressing toxic proteins [6]. Still, the downstream process (DSP), and especially the refolding unit operation, is the time-consuming step in gaining the correctly folded product from *E. coli* cultivations [28–31], which requires significantly more technology and time in purifying IBs [29,32,33]. Though IBs can be produced in such excess, the amount of generated product often outweighs the DSP efforts and makes the time-space yield more preferable for IBs [1,6,7,28].

One of the most favoured carbon sources in *E. coli* cultivations has always been glucose, as glucose has a very high affinity to the phosphotransferase system [34,35]. Glucose provides a lot of energy for the cells, as it is directly induced into glycolysis as glucose 6-phosphate and consumed through the tricarboxylic acid cycle (TCA) [35,36]. Usage of such, in combination with lactose, may result in diauxic growth and catabolite repression, which are caused by the regulatory network that is induced by glucose [37–39]. Catabolite repression results in decreased lactose uptake rates when glucose is present in excess [27,39,40]. Glycerol, first noticed in biotechnology as a by-product in the biodiesel production [41], has shown quite promising results in terms of biomass/substrate yield in *E. coli* cultivations [22,25]. To our knowledge, up to this point, no catabolic repression has been reported when glycerol was used as main carbon source (C-source) in combination with lactose [42]. In addition, mixtures of glucose, glycerol, and lactose have shown promising results for diverse products gained via autoinduction systems [20,25]. Recent studies [3,40] showed that the dependence of the inducer lactose influences the maximum IB production even on a quite low level of the specific glucose uptake rate. Low feeding rates of glucose would therefore result in the maximum inducer uptake rate, as cyclic adenosine monophosphate (cAMP) levels increase at higher glucose addition and therefore decrease the affinity for the RNA polymerase, decreasing the expression of the genes coding for the lac operon [35]. It is believed that cultivations with glycerol are able to overcome the problem of carbon catabolite repression and pave the way for usage of much higher specific C-source uptake rates, in order to increase time-space yields.

In this study, we performed cultivations with a *BL21(DE3)* strain, producing a recombinant protein coupled to a N-pro-fusion protein [43], expressed as IB with the goal to yield in maximum recombinant protein production. It is believed that glycerol causes positive results for the mixed-feed optimization when using lactose as an inducer, as glycerol—introduced into glycolysis but also into gluconeogenesis—yields a high amount of energy supplied to the cultivation system [42,44,45]. Couple that with increased cAMP levels throughout the whole cultivation, [35] glycerol is believed to be beneficial over a glucose cultivation system. It is shown that the recombinant protein production is increased compared to glucose, as a result of more available energy.

2. Materials and Methods

2.1. Bioreactor Cultivations

All cultivations were carried out with the strain *E. coli BL21(DE3)* consisting of the pET-30a plasmid system. The eukaryotic target protein was linked to a N-pro fusion taq (size of 28.8 kDA for the fusion protein) [43]. As the given protein is currently under patenting procedure at the industrial partner no detailed information can be given on the used protein.

All bioreactor and preculture cultivations were carried out using a defined minimal medium referred to DeLisa et al. (2015) [5]. Batch media and the preculture media had the same composition with different amounts of sugars respectively. The sugar concentrations for the phases are presented in Table 1:

Table 1. Respective sugar concentrations in media composition.

	Amount of Glucose	Amount of Glycerol
Preculture	8.8 g/L	8.9 g/L
Batch-Media	22 g/L	23 g/L
Feed	either 250 g/L or 300 g/L	

As pET-30a has a Kanamycin resistance gene, antibiotic was added throughout all fermentations, resulting in a final concentration of 0.02 g/L. All precultures were performed using 500 mL high yield flasks (containing the sugar concentrations given in Table 1). They were inoculated with 1.5 mL of bacteria solution stored in cryos at $-80\,^{\circ}C$ and subsequently cultivated for 20 h at 230 rpm in an Infors HR Multitron shaker (Infors, Bottmingen, Switzerland) at $37\,^{\circ}C$.

All cultivations were either performed in a DASGIP Mini bioreactor-4-parallel fermenter system (max. working volume: 2.5 L; Eppendorf, Hamburg, Germany) or in a DASbox Mini Bioreactor 4-parallel fermenter system (max. working V.: 250 mL; Eppendorf, Hamburg, Germany). For measuring the CO_2 and O_2 flows, a DASGIP-GA gas analyser was used (Eppendorf, Hamburg, Germany). The cultivations were controlled using the provided DAS-GIP-control system, DASware-control, which logged the process parameters. During cultivation, pH was kept constant at 7.2 and controlled with base only (12.5% NH_4OH), while acid (10% H_3PO_4) was added manually, if necessary. The pH was monitored using a pH-sensor EasyFerm Plus (Hamilton, Reno, NV, USA). Base addition was monitored observing the flowrates of a DASbox MP8 Multipumpmodul. The reactors were continuously stirred at 1400 rpm.

Aeration was absolved using mixture of pressurized air and pure oxygen at 2 vvm, mixing the ratios of the airflow, so that the dissolved oxygen (dO_2) was always higher than 40%. The dissolved oxygen was monitored using a fluorescence dissolved oxygen electrode Visiferm DO 120 (Hamilton, Reno, NV, USA).

2.2. Cultivation Scheme and q_s Screening Procedure

The batch media in the DASGIP reactors always contained 1 L DeLisa medium, while the DASbox Mini bioreactors contained a volume of 100 mL.

Only static q_s-controls were performed for these experiments, as the $q_{s,C}$ was not altered during induction phase [3,27]. The procedure was always as follows: Preculture, Batch, non-induced fed-batch, and induced fed batch with an adapted $q_{s,C}$.

Inoculation was always done with one tenth of the batch media volume, resulting in 100 mL of preculture. Preculture showed an OD_{600} of approximately 7 after cultivation (described above). The batch process, performed at $37\,^{\circ}C$, took around 6–7 h, depending on the C-source used, and was finished, visible by a drop in the CO_2 signal. The 22 g/L of either glucose or glycerol usually resulted in a biomass of 9–10 g/L. After the batch was finished, a non-induced fed-batch was started overnight, at 35 °C and adapting the $q_{s,C}$ value to gain a biomass of approximately 30 g/L. After the non-induced fed-batch, the volume was always decreased to 1 L, in order to keep induction conditions the same. Afterwards, $q_{s,C}$ was adapted to a certain point of interest, and temperature was decreased to 30 °C and stabilized for 30 min before the inducer was added. Induction was always performed with a lactose pulse of 100 mL of a 300 g/L sterile lactose solution—resulting in a lactose concentration in the fermentation broth of approximately 30 g/L. Induction period always lasted 7 h. The q_s control used here was performed using Equation (1) according to an exponential feed forward approach to keep q_s constant [3,27,40,46]:

$$F(t) = \frac{q_{s,\,C} \times X(t) \times \rho_f}{c_f} \qquad (1)$$

with F being the feed rate [g/h], $q_{s,C}$ the specific glucose or glycerol uptake rate [g/g/h], $X(t)$ the absolute biomass [g], ρ_f the feed density [g/L], and c_f the feed concentration [g/L], respectively.

2.3. Process Analytics

Samples are always taken after inoculation, upon end of the batch-phase and after the non-induced fed-batch was finished. During the induction period, samples were either taken in 20 or 30 min intervals. Generally, biomass was measured using OD_{600} and dry cell weight (DCW), while flow cytometry analysis (FCM) was used for determination of cell-death, especially in the induction phase. Optical density (OD_{600}) was measured using a Genesys 20 photometer (Thermo Scientific, Waltham, MA, USA). Since the linear range of the used photometer is between 0.2 and 0.8 [AU], samples were diluted with dH_2O to stay within that range. The dry cell weight was determined by vortexing the sample, pipetting 1 mL of sample solution in a pre-tared 2 mL Eppendorf-Safe-Lock Tube (Eppendorf, Hamburg, Germany), and centrifuged for 10 min at 11,000 rpm at 4 °C. After centrifugation, the supernatant was used immediately for at-line high-pressure liquid chromatography (HPLC) measurement (see beneath), while the pellet was re-suspended with 1 mL of 0.9% NaCl solution and centrifuged at the same conditions. Afterwards, the pellet was dried for at least 72 h at 105 °C. Samples for FCM were diluted 1:100 with 0.9% NaCl solution, stored at 4 °C, and measured after the process was finished. The measurement was performed using the software Cube 8 (Sysmex, Partec, Görlitz, Germany) according to Langemann et al. [47] using $DiBAC_4(3)$ (bis-(1,3-dibutylbarbituricacid) trimethineoxonol) and Rh414 dye. Rh414 binds to the plasma membrane and visualizes all cells, while DiBAC is sensitive to plasma membrane potential, and therefore distinction between viable and non-viable cells can be achieved.

Product samples were taken for [P]-strain, after 2, 5 and 7 h of induction phase. They were always treated as follows: 5 mL pipetted in a 50 mL Falcon tube, centrifuged for 10 min at 4800 rpm at 4 °C. The supernatant was discarded while the pellet was frozen at −20 °C. Samples for homogenisation were disrupted as follows: The pellets were re-suspended in a Lysis buffer (0.1 M TRIS, 10 mM EDTA, pH = 7.4) according to its dry cell weight (Equation (2)):

$$\text{Volume Lysis Puffer} = \text{DCW} \times \frac{5}{4} \qquad (2)$$

After suspending the cells, they were treated with an EmusiflexC3 Homogenizer (Avestin, Ottowa, ON, USA) at 1500 bar. The duration of homogenisation was always calculated to achieve ten passages through the homogenizer. After washing the pellets twice with dH_2O, the samples were measured using a HPLC method. The N-pro-fusion protein IB was measured via RP-HPLC (Thermo Scientific, Waltham, MA, USA) using a Nucleosil-column after solving in 7.5 M Guanidine Hydrochloride based buffer. The eluent was a gradient mixture of water with 0.1% TFA (tri-fluoric-acid) and Acetonitrile mixed with 0.1% TFA with a flow of 3 mL/min. Standard concentrations were 50, 140, 225, 320 and 500 mg/mL of an industrial supplied reference.

Sugar and glycerol concentrations were measured via HPLC-method (Thermo Scientific, Waltham, MA, USA) using a Supelcogel-column; Eluent: 0.1% H_3PO_4; Flow: 0.5 mL/min. Using this method, glucose or glycerol accumulation as well as the lactose decrease and the galactose accumulation could be detected. Standards had a concentration of 0.5, 1, 5, 10 and 20 g/L of every sugar used throughout all fermentations. The HPLC run lasted always for 25 min and chromatograms were analyzed using a Chromeleon Software (Dionex, Sunnyvale, CA, USA).

3. Results and Discussion

3.1. Mechanistic Correlations of Glycerol onto Specific Lactose Uptake Rate

The basic feeding rate for the induction phase for production of the recombined protein is a constant $q_{s,C}$—given by a fed-batch carried out on glucose or glycerol depending on the experiment—and by a pulse of 10 vol % high concentrated lactose feed.

In order to get comparable datasets for all experiments, a mechanistic model approach is performed. As shown in previous studies, the maximum possible specific lactose uptake rates depend on the specific glucose uptake rates which can be described by a mechanistic equation (see Equation (3)) [3,40]. The maximum $q_{s,lac}$ rates depend Monod-like on $q_{s,glu}$ until a certain maximum is reached at a respectively low feeding rate of glucose, before $q_{s,lac}$ decreases at high $q_{s,glu}$ which performs analogue to substrate inhibition [3]. Values for y = 0 correspond to the uptake rates on sole glucose/glycerol, respectively.

$$q_{s,lac} = q_{s,lac,max} \times \max\left(\left(1 - \frac{q_{s,glu}}{q_{s,glu,crit}}\right)^n, 0\right) \times \left(\frac{q_{s,glu}}{q_{s,glu} + K_A} + \frac{q_{s,lac,noglu}}{q_{s,lac,\,max}}\right) \qquad (3)$$

with $q_{s,lac}$ being the specific lactose uptake rate [g/g/h], $q_{s,lac,max}$ the maximum specific lactose uptake rate [g/g/h], $q_{s,glu}$ the specific glucose uptake rate [g/g/h], $q_{s,glu,crit}$ the critical specific glucose uptake rate up to which lactose is consumed [g/g/h], $q_{s,lac,noglu}$ the specific lactose uptake rate at $q_{s,glu} = 0$ [g/g/h], and K_A the affinity constant for the specific lactose uptake rate [g/g/h]. n describes the type of inhibition (non-competitive, uncompetitive, competitive).

As the model has already been established for four different products in glucose-lactose systems [40], it had to be shown if the same function fits for the given product. We fitted the model parameters as described in Wurm et al., where also a detailed description of the model derivation can be found [3]. As shown in Figure 1 and Table 2, parameters can be found to describe the experimental data for glucose and glycerol as C-source. In absence of glucose, lactose cannot be taken up, since there is not enough adenosine triphosphate (ATP) produced. Once a certain threshold of glucose is passed, enough ATP is created to metabolize the inducer [3,40]. The trend seen in the cultivations performed on glucose are explained by the well-known phenomenon of catabolite repression (CCR) [37,39], as the lactose uptake rates decrease significantly with increasing the feeding rate. As *E. coli* BL21(DE3) is not able to metabolize galactose due to absence of a (gal) gene, which can be referred to a deletion of the genes gal M, K, T, E [48,49], galactose should accumulate in the fermentation broth [37,50]. Hence, the galactose accumulation rate in the fermentation broth could generally be correlated to the lactose depletion rate during the cultivation (not shown).

However, the curves for glucose and glycerol are almost identical. Generally, a higher affinity for glucose is reported in literature [35], resulting in a higher μ for those cultivations, as glycerol has less affinity to the phosphotransferase system (PTS) [37]. This trend is in accordance with our data given in the value $q_{s,C,crit}$ in Table 2. Furthermore, biomass to substrate yields ($Y_{X/S}$) for glucose decrease in the induction phase from about 0.5 in the batch phase to about 0.336 ± 0.05 after the one-point lactose addition. By contrast, $Y_{X/S}$ of glycerol are generally about 0.44 ± 0.1 during the induction phase [51].

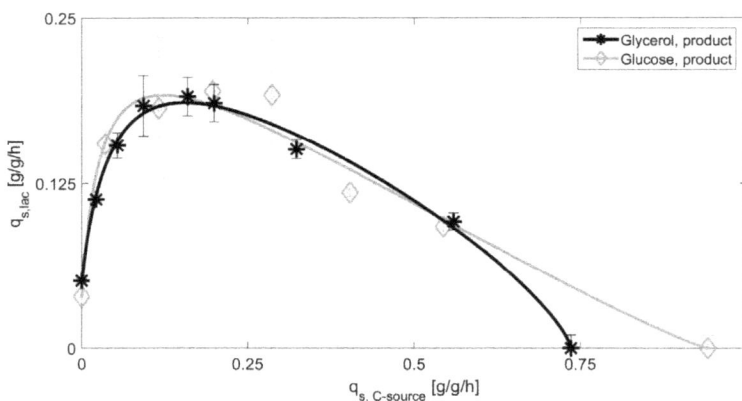

Figure 1. Extracted datapoints for q_s values including standard deviations for cultivations with glucose and glycerol in the production strain (glycerol product, glucose product). Solid lines represent the model based approach for inducer uptake rates vs. feeding rates models of glucose and glycerol.

Table 2. Model parameters and normalized-root-mean-square-error (NRMSE) for the different analysed cultivation with produced product (P).

Cultivation System	$q_{s,lac,max}$	K_A	$q_{s,C,crit}$	n	$q_{s,lac,noglu}$	NRMSE
	[g/g/h]	[g/g/h]	[g/g/h]	[-]	[g/g/h]	[%]
Glucose	0.23	0.032	0.94	1.14	0.039	6.5
Glycerol	0.23	0.053	0.74	0.74	0.051	2.6

This does not explain the very similar lactose uptake values at high $q_{s,C}$, since it is believed that carbon catabolite repression should not be present using glycerol as primary carbon source [52]. The production of the recombinant protein seems to induce stress resulting in the maximum possible activity inside the cell, which is represented by the similarity of the two curves. Therefore, the decrease of the $q_{s,lac}$ rate in the model-based approach actually referred to the CCR for glucose based systems so far ($\left(1 - \frac{q_{s,glu}}{q_{s,glu,crit}}\right)^n$), may have to be reconsidered when glycerol is fed. In turn, our results would indicate that the decline cannot be attributed to carbon catabolite repression, also not for glucose. Glycerol does not interfere with the PTS transport system and no resulting change of the cAMP levels during uptake of lactose are to be believed on a first glance. Glycerol enters glycolysis as di-hydroxy-acetone-phosphate and is processed in glycolysis producing pyruvate, but also there are gluconeogenetic genes active providing the formation of glucose-6-phosphate [41,53,54]. As glycolysis seems to be running at maximum capacity, a bottleneck in the trycarboxylic acid (TCA) cycle may also be likely. Overload of the TCA cycle has already been described by Heyland et al. (2011) [55], saying that the TCA cycle cannot metabolize all the pyruvate produced in glycolysis. It has also been referred that the cells try to gain energy in alternative ways such as using acetate as a terminal electron acceptor, or the usage of oxidative phosphorylation [55,56]. However, as *E. coli BL21(DE3)* produces relatively low levels of acetate in general, the acetate formation is always beneath the threshold of the HPLC and may therefore not the predominant electron acceptor in this strain.

To test the observed effects, we tried a process technological method approach, rather than performing expensive and time consuming "omics" analysis. The pET-30a plasmid was transformed into the used strain *E. coli BL21(DE3)* without the sequence for the recombinant protein, further referred as non-producer (NP) strain. The strain was tested in the same analytical way as the used strain for recombinant protein production. HPLC raw data for lactose decrease are compared with an almost identical $q_{s,C}$ (~0,1 g/g/h) in Figure 2.

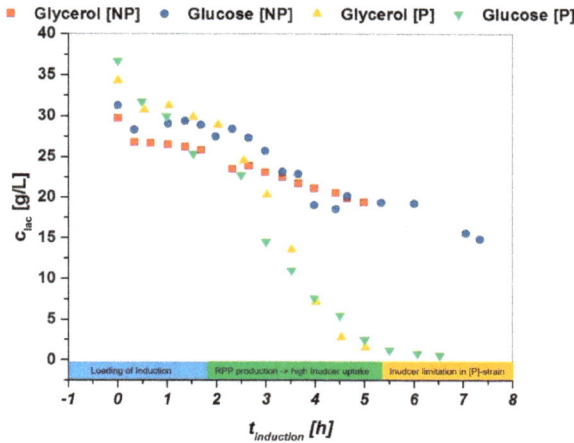

Figure 2. High-pressure liquid chromatography (HPLC)-based data for decrease of lactose in fermentation broth exhibiting very similar $q_{s,C}$ values in [g/L]. A significant decrease over the time of induction is visible in producing (P) strains, while the decrease is way slower in non-producing (NP)-strain-cultivations.

Hereby, three phases can be seen for the product producing strain in the induction phase, while only two phases can be seen in the NP strain:

(i)   Adaption phase: lactose gets transferred to alloactose and loads the induction (0–2 h in induction phase).
(ii) Linear decrease of lactose as the system needs inducer for recombinant protein expression (2–5 h).
(iii) Limitation of lactose in P strain: not sufficient inducer present, need for mixed feed system (5–7 h), no inducer limitation seen in NP strain, further decrease of inducer analogue to phase 2.

Results on the model-based approach for the glucose system are given in Figure 3.

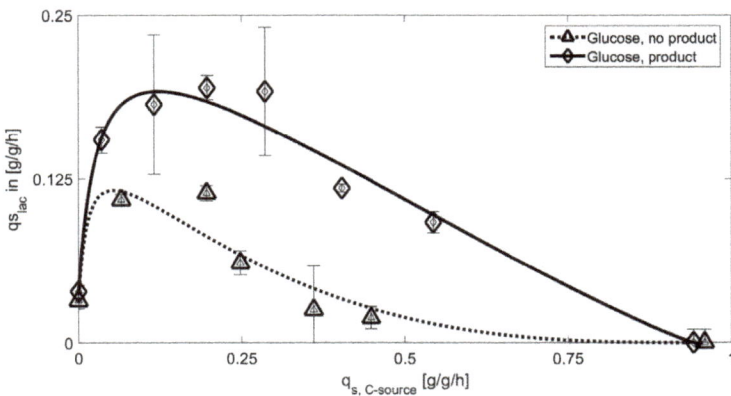

Figure 3. Extracted datapoints for $q_{s,C}$ values including standard deviations for cultivations with glucose using the product producing (glucose product) and the NP strain (glucose, no product). Solid lines represent the model based approach for inducer uptake rates vs. feeding rates models of glucose. A clearly visible difference can be observed during these cultivations.

The fermentations performed with the NP-strain showed lactose uptake rates resemble the expected carbon catabolite repression for glucose including high affinity of the PTS system at low $q_{s,glu}$ which can also be seen in Table 3. Despite the identical behavior of protein producing and NP strain, a clear difference in maximum $q_{s,lac}$ is obviously present. Higher consumption of glucose has impact on the cAMP level and decreases the specific uptake of lactose in the product producing strain. $Y_{X/S}$ stays very similar in both cases 0.37 ± 0.05 for the protein producing strain vs. 0.383 ± 0.053 for the NP strain. Given yields are a mean value over all $q_{s,C}$ values except for (lac) = 0 and (glu) = 0. So, these general deviations in $q_{s,lac,max}$ can be attributed to the increased energy demand during recombinant product production, as also the biomass yields stay the same. Lactose uptake rates on glycerol for the product producing and the NP strain are given in Figure 4. Despite the quite straightforward mechanistic explanation for glucose, glycerol biomass to substrate yields differ fundamentally for both experiments: $Y_{X/S} = 0.55 \pm 0.11$ for the NP strain, while the producing strain has a $Y_{X/S}$ of 0.44 ± 0.1. This fact may explain the much shallower uptake at low $q_{s,C}$ for the NP strain, but cannot explain the difference in the CCR term.

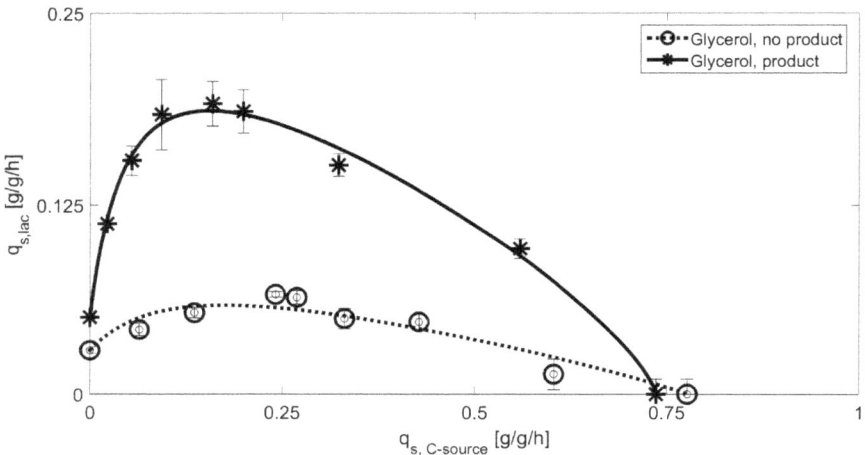

Figure 4. Extracted datapoints for $q_{s,C}$ values including standard deviations for cultivations with glycerol using the product producing (glycerol, product) and the NP strain (glycerol, no product). Solid lines represent the model based approach for inducer uptake rates vs. feeding rates models of glucose.

As a far higher biomass yield is present in the NP strain, only a reduced amount of lactose is taken up, which explains the decreased $q_{s,lac,max}$. However, the NP strain shows no pronounced substrate inhibition. The carbon catabolite repression term of the model on glycerol has only low impact (see Table 3), as the upregulation of cAMP using glycerol would also be beneficial for the lactose uptake mechanism in the PTS system [35]. Since the lactose facilitator is not considered to be the rate determining step in the glycerol metabolism, glycerol kinase closely regulated to the PTS system may cause the CCR-like effects [44,45]. As the feeding rate increases, the possibility of short-term local glucose and glycerol accumulation increases, eventually leading to diauxic growth and therefore decreased lactose rates as glucose and glycerol have higher affinity than disaccharides for *E. coli* [35,52,57,58]. The product-producing strain shows a high regulated lactose uptake at low $q_{s,C}$ values, as a result of lower biomass yield and higher energy demand in production of the recombinant protein. Higher lactose uptake results in high intracellular glucose level, which show the similar feedback mechanism like in the glucose fed cultivations.

Table 3. Model parameters and normalized-root-mean-square-error (NRMSE) for the analysed cultivation without recombinant product production (NP).

Cultivation System	$q_{s,lac,max}$ [g/g/h]	K_A [g/g/h]	$q_{s,glu,crit}$ [g/g/h]	n [-]	$q_{s,lac,noglu}$ [g/g/h]	NRMSE [%]
Glucose [NP]	0.14	0.016	0.96	2.92	0.032	12.7
Glycerol [NP]	0.10	0.13	0.78	0.90	0.029	9.7

As a result, both curves given in Figure 1 have a very similar appearance, but are expected to have a very different regulation within. To get insight into respiratory activity, qCO_2 values are compared for all four fermentations, respectively. Evaluation is given in Table 4 based on the applied $q_{s,C}$ values.

Table 4. Specific substrate uptake rate vs. specific carbon evolution rate. Product producing strains have in general increased respiratory activity. NP strains show reduced respiratory activity. Standard deviation of qCO_2 increases at higher feeding rates.

Glucose		Glucose NP		Glycerol		Glycerol NP	
$q_{s,C}$ [g/g/h]	qCO_2 [g/g/h]	$q_{s,C}$ [g/g/h]	qCO_2 [g/g/h]	$q_{s,C}$ [g/g/h]	qCO_2 [g/g/h]	$q_{s,C}$ [g/g/h]	qCO_2 [g/g/h]
0.036	2.15 ± 0.33	0.066	1.69 ± 0.25	0.022	2.91 ± 0.46	0.064	0.82 ± 0.09
0.116	3.12 ± 0.46	0.196	3.75 ± 0.44	0.054	4.41 ± 0.78	0.136	1.85 ± 0.21
0.197	3.98 ± 0.55	0.224	3.35 ± 0.42	0.093	3.88 ± 0.64	0.225	2.86 ± 0.31
0.286	5.72 ± 0.41	0.36	5.96 ± 0.26	0.159	3.12 ± 0.43	0.331	3.31 ± 0.22
0.403	6.42 ± 1.48	0.448	5.64 ± 0.47	0.199	4.14 ± 0.64	0.428	4.07 ± 0.51
0.544	7.30 ± 1.64			0.323	5.13 ± 0.48	0.603	1.75 ± 1.58
				0.559	7.18 ± 2.10		

Highly similar respiratory activity is received for the product producing strain, almost linear increasing with $q_{s,C}$. For the NP strain, a general lower respiratory activity is seen for the glycerol-fed strain. These results support the fact that lower energy demand is needed in this strain based on the general higher biomass yield and the fact that no recombinant protein is produced. In TCA, first steps of amino acid synthesis are performed, therefore the production of non-essential AA would result in the accumulation of NADH [59]. As approximately two NADH molecules can be formed to one molecule of CO_2 the enhanced respiratory activity in the product producing strain is most likely coding for the enhanced production of non-essential AA, which are essential for the recombinant product. However, further analysis on stress induced changes in the gene expression may give valuable new insights into regulation mechanism in *E. coli*.

3.2. Productivity and Physiology Using Glycerol as Primary Carbon Source

As the overall goal is an increased production rate of recombinant protein, we compare titers of the produced IBs as a function of carbon source and uptake rate. In Figure 5a, the increase in IB titer over time is presented for two cultivations. The loading of the induction, which takes about 2 h, can be clearly dedicated in these results, with no titer of the recombinant protein to be found within the first 2 h (also compare to Figure 2). Figure 5b shows product IB titers after 7 h induction time, which are plotted against the corresponding $q_{s,C}$. Only the feed rate of glucose/glycerol, adapted for the static experiment in the induced fed-batch phase, is used in this plot—as cultivations are induced with one lactose pulse only, the $q_{s,C}$ is a non-cumulative one. Generally, an increase in the feeding rate is beneficial for product formation. Cultivations carried out on glycerol tend to produce more recombinant protein with a product optimum at a q_s-glycerol-level seen around 0.3–0.35 g/g/h. It may be possible that even higher product titers can be found within the range of 0.3–0.55 g/g/h. Cultivations carried out on glucose also tend to produce more product when the feeding rate is shifted to rather high rates as well. Very similar IB titers can be obtained at high $q_{s,C}$ levels, but are far away from the observed maximum.

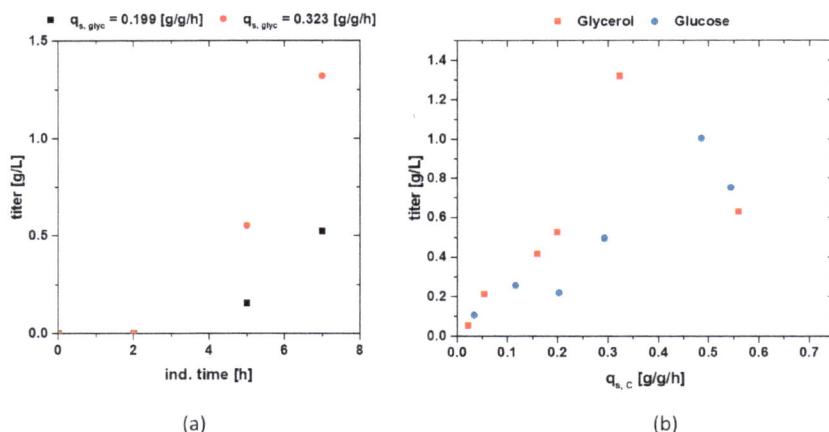

Figure 5. (**a**) Time dependence for two Inclusion Body (IB) titers starting from lactose addition to 7 h of induction; (**b**) Titers of the recombinant produced protein, after homogenisation of the inclusion bodies and a two-time washing plotted vs. the q_s of glucose and glycerol; A trend can be seen in gaining more product when cultivations are carried out on glycerol compared to glucose, respectively.

The high increase in titer as a function of $q_{s,C}$ in glycerol may be a result of the higher biomass (higher $Y_{X/S}$ during induction) usually present in glycerol fed induction phases. The phenomenon of high product formation rates at high feeding levels, was much to our surprise, as we expected to see enhanced stress reactions by the cells due to overfeeding—especially at later time stages—usually present in IPTG induced cultures. Though we see only very little levels of glucose or glycerol accumulation in our HPLC measurements (data not shown). This could be, as the fermentation conditions in the induction phase are respectively mild. Temperature is decreased to 30 °C and induction with lactose is regarded to be a softer induction than IPTG, as lactose can be metabolized by *E. coli* [22,23]. In literature, it has been reported that the catabolic repression increases with higher temperatures [60]. Altering the temperature in the induction phase would have probably led to very different results in lactose uptake rates as well as different product data. Also, we want to highlight that every induction here was only performed with a one-time lactose pulse, which is most likely an insufficient induction, as there may be too little inducer in the media, which can be seen in Figure 2. In the following development steps, mixed feeds using glycerol in combination with lactose must be established and measured as this would lead to a constant and complete induction of the system. However, the product data supports the results that most probably very different regulation mechanisms in *E. coli* lead to the same visible uptake rates in Figure 1, but have severe effects on the productivity on the different carbon sources.

Physiological analysis using flow cytometry (FCM) is presented in Figure 6a,b. The NP strain given in Figure 6a has very similar appearance for glucose and glycerol, respectively, increasing number of dead cells by increasing the feeding rate beyond a certain threshold, imposing stress to the cell. Throughout the whole experimental design, producing cells grown on glycerol exhibit a smaller cell size compared to cells grown on glucose (not shown). Since cell debris and residual particles are seen at similar cell sizes like glycerol grown cells a general higher abundance is present during those cultivations. To cope with this problem, FCM data after the non-induced fed-batch is subtracted from the subsequent measurements.

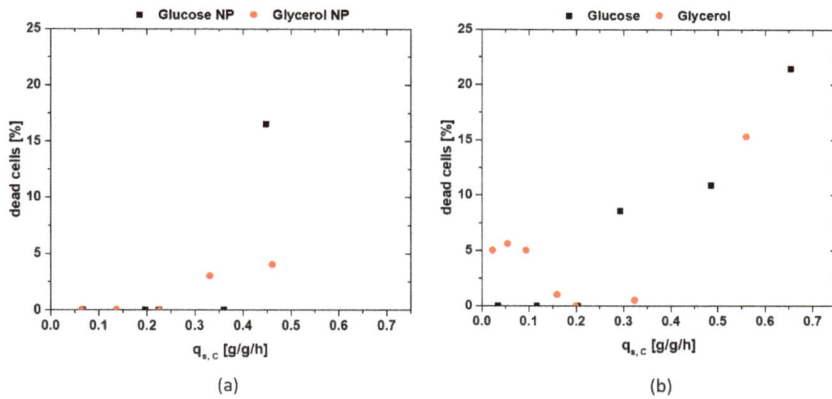

Figure 6. (a) Flow cytometry (FCM) analysis of NP strain 5 h after lactose pulse. As no protein data are received from these cultivations, the induction time was limited to 5 h; (b) FCM analysis of the product producing strain. Glycerol imposes stress at low feeding rates, while glucose shows increase in cell stress beginning at about 0.25 g/g/h.

The viability of both cultivation strategies for an induction time of 6 h—often used for IB production at industrial scale—is given in Figure 6b, with a strong contrast between glucose and glycerol. While cells fed with glucose show no cell lysis at low $q_{s,C}$ levels and are very similar to NP strain in Figure 6a, glycerol shows certain stress reaction resulting in about 5% dead cells until a 0.2 g/g/h. Afterwards, stable conditions for glycerol can be found, while stress is induced at glucose-fed systems starting at about 0.25 g/g/h. As the overnight fed-batch phase generally exhibited a q_s of 0.25 g/g/h, the switch to very low $q_{s,C}$ in the induction phase, combined with the lactose pulse, may impose the cell stress seen in 5% dead cells in Figure 6b. This corresponds well to the product data in Figure 5 with similar or even higher productivity of glucose at low $q_{s,C}$ levels, but higher productivity for glycerol at moderate to high levels. Including the fact that glycerol shows higher biomass yields during induction with lactose, glycerol may be well used as an alternative main carbon source in *E. coli* cultivations, even though glucose has high affinity to the phosphotransferase system (PTS). It has already been reported that addition of glycerol to a glucose-lactose induction system increases product formation [20,61]. As glycerol needs increased cAMP levels, which are also needed for lactose uptake [37], this might be a key function in regulating higher lactose uptake and subsequently increasing productivity and product titer.

Furthermore, as glycerol is a cheap media compared to glucose, an application of glycerol in mixed-feed system with lactose may be highly beneficial for recombinant protein production performed in industry.

4. Conclusions

In this work, the effects of glycerol or glucose on lactose uptake rates for an IB-based process using *E. coli* BL21(DE3) were investigated. Feeding and uptake rates are compared and evaluated in terms of productivity and physiology using FCM.

It is shown that both C-sources show identical lactose uptake rates as a function of $q_{s,C}$. The used model-based approach already performed for different products in Wurm et al. [40] can be used for description of both curves. It has been detected that glycerol is beneficial over the usage of glucose for maximising the recombinant protein production of a lactose induced system.

Glycerol and glucose most probably exhibit different regulation of the carbon catabolite repression—the reduction of lactose uptake at higher $q_{s,C}$ levels. This hypothesis is supported by cultivation and evaluation of a non-producer strain exhibiting the expected behaviour for both

C-sources, respectively. As this behaviour was not seen in the producing strain, it seems like the expression hosts are performing at maximum capacity in recombinant protein production. Additionally, glycerol is referred to different metabolic pathways [42], eventually increasing the metabolic flux [55] towards recombinant protein production.

Physiology and productivity support the hypothesis that glycerol is promising C-source for cultivations using mixed feed systems with moderate to high $q_{s,C}$ values in order to boost time-space yields. As scale-up in *E. coli* systems can be performed relatively easily [1], the much lower costs of glycerol, when compared to glucose respectively, might provide interesting options for industrial and other large scale applications.

Acknowledgments: We gratefully thank the Christian Doppler Society for the funding of this work.

Author Contributions: Julian Kopp performed and cultivations, and calculated the q_s-q_s-curves, Christoph Slouka performed the purification and the analysis of product data (These authors contributed equally to this work). Julian Kager supported during the cultivations. Sophia Ulonska is responsible for the model evaluation. Jens Fricke, Oliver Spadiut and Christoph Herwig gave valuable information for drafting this paper and helped during preparation.

Conflicts of Interest: The authors declare no conflict of interest.

References

1. Gupta, S.K.; Shukla, P. Microbial platform technology for recombinant antibody fragment production: A review. *Crit. Rev. Microbiol.* **2017**, *43*, 31–42. [CrossRef] [PubMed]
2. Walsh, G. Biopharmaceutical benchmarks 2010. *Nat. Biotechnol.* **2010**, *28*, 917–924. [CrossRef] [PubMed]
3. Wurm, D.J.; Veiter, L.; Ulonska, S.; Eggenreich, B.; Herwig, C.; Spadiut, O. The *E. coli* pET expression system revisited-mechanistic correlation between glucose and lactose uptake. *Appl. Microbiol. Biotechnol.* **2016**, *100*, 8721–8729. [CrossRef] [PubMed]
4. Meuris, L.; Santens, F.; Elson, G.; Festjens, N.; Boone, M.; Dos Santos, A.; Devos, S.; Rousseau, F.; Plets, E.; Houthuys, E.; et al. GlycoDelete engineering of mammalian cells simplifies N-glycosylation of recombinant proteins. *Nat. Biotechnol.* **2014**, *32*, 485–489. [CrossRef] [PubMed]
5. DeLisa, M.P.; Li, J.; Rao, G.; Weigand, W.A.; Bentley, W.E. Monitoring GFP-operon fusion protein expression during high cell density cultivation of Escherichia coli using an on-line optical sensor. *Biotechnol. Bioeng.* **1999**, *65*, 54–64. [CrossRef]
6. Berlec, A.; Strukelj, B. Current state and recent advances in biopharmaceutical production in *Escherichia coli*, yeasts and mammalian cells. *J. Ind. Microbiol. Biotechnol.* **2013**, *40*, 257–274. [CrossRef] [PubMed]
7. Baeshen, M.N.; Al-Hejin, A.M.; Bora, R.S.; Ahmed, M.M.; Ramadan, H.A.; Saini, K.S.; Baeshen, N.A.; Redwan, E.M. Production of Biopharmaceuticals in *E. coli*: Current Scenario and Future Perspectives. *J. Microbiol. Biotechnol.* **2015**, *25*, 953–962. [CrossRef] [PubMed]
8. Spadiut, O.; Capone, S.; Krainer, F.; Glieder, A.; Herwig, C. Microbials for the production of monoclonal antibodies and antibody fragments. *Trends Biotechnol.* **2014**, *32*, 54–60. [CrossRef] [PubMed]
9. Studier, F.W.; Moffatt, B.A. Use of bacteriophage T7 RNA polymerase to direct selective high-level expression of cloned genes. *J. Mol. Biol.* **1986**, *189*, 113–130. [CrossRef]
10. Steen, R.; Dahlberg, A.E.; Lade, B.N.; Studier, F.W.; Dunn, J.J. T7 RNA polymerase directed expression of the Escherichia coli rrnB operon. *EMBO J.* **1986**, *5*, 1099–1103. [PubMed]
11. Studier, F.W.; Rosenberg, A.H.; Dunn, J.J.; Dubendorff, J.W. Use of T7 RNA polymerase to direct expression of cloned genes. *Methods Enzymol.* **1990**, *185*, 60–89. [PubMed]
12. Dubendorff, J.W.; Studier, F.W. Controlling basal expression in an inducible T7 expression system by blocking the target T7 promoter with lac repressor. *J. Mol. Biol.* **1991**, *219*, 45–59. [CrossRef]
13. Neubauer, P.; Hofmann, K. Efficient use of lactose for the lac promoter-controlled overexpression of the main antigenic protein of the foot and mouth disease virus in Escherichia coli under fed-batch fermentation conditions. *FEMS Microbiol. Rev.* **1994**, *14*, 99–102. [CrossRef] [PubMed]
14. Lyakhov, D.L.; He, B.; Zhang, X.; Studier, F.W.; Dunn, J.J.; McAllister, W.T. Pausing and termination by bacteriophage T7 RNA polymerase. *J. Mol. Biol.* **1998**, *280*, 201–213. [CrossRef] [PubMed]

15. Jeong, H.; Barbe, V.; Lee, C.H.; Vallenet, D.; Yu, D.S.; Choi, S.H.; Couloux, A.; Lee, S.W.; Yoon, S.H.; Cattolico, L.; et al. Genome sequences of *Escherichia coli* B strains REL606 and BL21(DE3). *J. Mol. Biol.* **2009**, *394*, 644–652. [CrossRef] [PubMed]

16. Jeong, H.; Kim, H.J.; Lee, S.J. Complete Genome Sequence of *Escherichia coli* Strain BL21. *Genome Announc.* **2015**, *3*, e00134-15. [CrossRef] [PubMed]

17. Tseng, T.T.; Tyler, B.M.; Setubal, J.C. Protein secretion systems in bacterial-host associations, and their description in the Gene Ontology. *BMC Microbiol.* **2009**, *9* (Suppl. 1), S2. [CrossRef] [PubMed]

18. Marbach, A.; Bettenbrock, K. Lac operon induction in *Escherichia coli*: Systematic comparison of IPTG and TMG induction and influence of the transacetylase LacA. *J. Biotechnol.* **2012**, *157*, 82–88. [CrossRef] [PubMed]

19. Keiler, K.C. Biology of trans-translation. *Annu. Rev. Microbiol.* **2008**, *62*, 133–151. [CrossRef] [PubMed]

20. Viitanen, M.I.; Vasala, A.; Neubauer, P.; Alatossava, T. Cheese whey-induced high-cell-density production of recombinant proteins in *Escherichia coli*. *Microb. Cell Fact.* **2003**, *2*, 2. [CrossRef] [PubMed]

21. Marisch, K.; Bayer, K.; Cserjan-Puschmann, M.; Luchner, M.; Striedner, G. Evaluation of three industrial *Escherichia coli* strains in fed-batch cultivations during high-level SOD protein production. *Microb. Cell Fact.* **2013**, *12*, 58. [CrossRef] [PubMed]

22. Ukkonen, K.; Mayer, S.; Vasala, A.; Neubauer, P. Use of slow glucose feeding as supporting carbon source in lactose autoinduction medium improves the robustness of protein expression at different aeration conditions. *Protein Expr. Purif.* **2013**, *91*, 147–154. [CrossRef] [PubMed]

23. Neubauer, P.; Hofmann, K.; Holst, O.; Mattiasson, B.; Kruschke, P. Maximizing the expression of a recombinant gene in *Escherichia coli* by manipulation of induction time using lactose as inducer. *Appl. Microbiol. Biotechnol.* **1992**, *36*, 739–744. [CrossRef] [PubMed]

24. Marschall, L.; Sagmeister, P.; Herwig, C. Tunable recombinant protein expression in *E. coli*: Enabler for continuous processing? *Appl. Microbiol. Biotechnol.* **2016**, *100*, 5719–5728. [CrossRef] [PubMed]

25. Blommel, P.G.; Becker, K.J.; Duvnjak, P.; Fox, B.G. Enhanced bacterial protein expression during auto-induction obtained by alteration of lac repressor dosage and medium composition. *Biotechnol. Prog.* **2007**, *23*, 585–598. [CrossRef] [PubMed]

26. Dvorak, P.; Chrast, L.; Nikel, P.I.; Fedr, R.; Soucek, K.; Sedlackova, M.; Chaloupkova, R.; de Lorenzo, V.; Prokop, Z.; Damborsky, J. Exacerbation of substrate toxicity by IPTG in *Escherichia coli* BL21(DE3) carrying a synthetic metabolic pathway. *Microb. Cell Fact.* **2015**, *14*, 201. [CrossRef] [PubMed]

27. Wurm, D.J.; Herwig, C.; Spadiut, O. How to Determine Interdependencies of Glucose and Lactose Uptake Rates for Heterologous Protein Production with *E. coli*. *Methods Mol. Biol.* **2017**, *1586*, 397–408. [PubMed]

28. García-Fruitós, E.; Vázquez, E.; Díez-Gil, C.; Corchero, J.L.; Seras-Franzoso, J.; Ratera, I.; Veciana, J.; Villaverde, A. Bacterial inclusion bodies: Making gold from waste. *Trends Biotechnol.* **2012**, *30*, 65–70. [CrossRef] [PubMed]

29. Palmer, I.; Wingfield, P.T. Preparation and extraction of insoluble (inclusion-body) proteins from *Escherichia coli*. *Curr. Protoc. Protein Sci.* **2012**. [CrossRef]

30. Ramón, A.; Señorale-Pose, M.; Marín, M. Inclusion bodies: Not that bad . . . . *Front. Microbiol.* **2014**, *5*, 56. [CrossRef] [PubMed]

31. Villaverde, A.; Corchero, J.L.; Seras-Franzoso, J.; Garcia-Fruitós, E. Functional protein aggregates: Just the tip of the iceberg. *Nanomedicine (Lond.)* **2015**, *10*, 2881–2891. [CrossRef] [PubMed]

32. Wingfield, P.T.; Palmer, I.; Liang, S.M. Folding and Purification of Insoluble (Inclusion Body) Proteins from *Escherichia coli*. *Curr. Protoc. Protein Sci.* **2014**. [CrossRef]

33. Wingfield, P.T. Preparation of Soluble Proteins from *Escherichia coli*. *Curr. Protoc. Protein Sci.* **2014**, *78*, 6.2.1–6.2.22. [PubMed]

34. Postma, P.W.; Lengeler, J.W.; Jacobson, G.R. Phosphoenolpyruvate:carbohydrate phosphotransferase systems of bacteria. *Microbiol. Rev.* **1993**, *57*, 543–594. [PubMed]

35. Deutscher, J.; Francke, C.; Postma, P.W. How phosphotransferase system-related protein phosphorylation regulates carbohydrate metabolism in bacteria. *Microbiol. Mol. Biol. Rev.* **2006**, *70*, 939–1031. [CrossRef] [PubMed]

36. Ronimus, R.S.; Morgan, H.W. Distribution and phylogenies of enzymes of the Embden-Meyerhof-Parnas pathway from archaea and hyperthermophilic bacteria support a gluconeogenic origin of metabolism. *Archaea* **2003**, *1*, 199–221. [CrossRef] [PubMed]

37. Bettenbrock, K.; Fischer, S.; Kremling, A.; Jahreis, K.; Sauter, T.; Gilles, E.D. A quantitative approach to catabolite repression in *Escherichia coli*. *J. Biol. Chem.* **2006**, *281*, 2578–2584. [CrossRef] [PubMed]
38. Kremling, A.; Bettenbrock, K.; Laube, B.; Jahreis, K.; Lengeler, J.W.; Gilles, E.D. The organization of metabolic reaction networks. III. Application for diauxic growth on glucose and lactose. *Metab. Eng.* **2001**, *3*, 362–379. [CrossRef] [PubMed]
39. Stülke, J.; Hillen, W. Carbon catabolite repression in bacteria. *Curr. Opin. Microbiol.* **1999**, *2*, 195–201. [CrossRef]
40. Wurm, D.J.; Hausjell, J.; Ulonska, S.; Herwig, C.; Spadiut, O. Mechanistic platform knowledge of concomitant sugar uptake in *Escherichia coli* BL21(DE3) strains. *Sci. Rep.* **2017**, *7*, 45072. [CrossRef] [PubMed]
41. Martínez-Gómez, K.; Flores, N.; Castañeda, H.M.; Martínez-Batallar, G.; Hernández-Chávez, G.; Ramírez, O.T.; Gosset, G.; Encarnación, S.; Bolivar, F. New insights into *Escherichia coli* metabolism: Carbon scavenging, acetate metabolism and carbon recycling responses during growth on glycerol. *Microb. Cell Fact.* **2012**, *11*, 46. [CrossRef] [PubMed]
42. Lin, E.C. Glycerol dissimilation and its regulation in bacteria. *Annu. Rev. Microbiol.* **1976**, *30*, 535–578. [CrossRef] [PubMed]
43. Achmüller, C.; Kaar, W.; Ahrer, K.; Wechner, P.; Hahn, R.; Werther, F.; Schmidinger, H.; Cserjan-Puschmann, M.; Clementschitsch, F.; Striedner, G.; et al. N(pro) fusion technology to produce proteins with authentic N termini in *E. coli*. *Nat. Methods* **2007**, *4*, 1037–1043. [CrossRef] [PubMed]
44. Zwaig, N.; Kistler, W.S.; Lin, E.C. Glycerol kinase, the pacemaker for the dissimilation of glycerol in *Escherichia coli*. *J. Bacteriol.* **1970**, *102*, 753–759. [PubMed]
45. Voegele, R.T.; Sweet, G.D.; Boos, W. Glycerol kinase of *Escherichia coli* is activated by interaction with the glycerol facilitator. *J. Bacteriol.* **1993**, *175*, 1087–1094. [CrossRef] [PubMed]
46. Slouka, C.; Wurm, D.J.; Brunauer, G.; Welzl-Wachter, A.; Spadiut, O.; Fleig, J.; Herwig, C. A Novel Application for Low Frequency Electrochemical Impedance Spectroscopy as an Online Process Monitoring Tool for Viable Cell Concentrations. *Sensors (Basel)* **2016**, *16*, 1900. [CrossRef] [PubMed]
47. Langemann, T.; Mayr, U.B.; Meitz, A.; Lubitz, W.; Herwig, C. Multi-parameter flow cytometry as a process analytical technology (PAT) approach for the assessment of bacterial ghost production. *Appl. Microbiol. Biotechnol.* **2016**, *100*, 409–418. [CrossRef] [PubMed]
48. Xu, J.; Banerjee, A.; Pan, S.H.; Li, Z.J. Galactose can be an inducer for production of therapeutic proteins by auto-induction using *E. coli* BL21 strains. *Protein Expr. Purif.* **2012**, *83*, 30–36. [CrossRef] [PubMed]
49. Studier, F.W.; Daegelen, P.; Lenski, R.E.; Maslov, S.; Kim, J.F. Understanding the differences between genome sequences of *Escherichia coli* B strains REL606 and BL21(DE3) and comparison of the *E. coli* B and K-12 genomes. *J. Mol. Biol.* **2009**, *394*, 653–680. [CrossRef] [PubMed]
50. Daegelen, P.; Studier, F.W.; Lenski, R.E.; Cure, S.; Kim, J.F. Tracing ancestors and relatives of *Escherichia coli* B, and the derivation of B strains REL606 and BL21(DE3). *J. Mol. Biol.* **2009**, *394*, 634–643. [CrossRef] [PubMed]
51. Korz, D.J.; Rinas, U.; Hellmuth, K.; Sanders, E.A.; Deckwer, W.D. Simple fed-batch technique for high cell density cultivation of *Escherichia coli*. *J. Biotechnol.* **1995**, *39*, 59–65. [CrossRef]
52. Inada, T.; Kimata, K.; Aiba, H. Mechanism responsible for glucose-lactose diauxie in *Escherichia coli*: Challenge to the cAMP model. *Genes Cells* **1996**, *1*, 293–301. [CrossRef] [PubMed]
53. Larson, T.J.; Ye, S.Z.; Weissenborn, D.L.; Hoffmann, H.J.; Schweizer, H. Purification and characterization of the repressor for the sn-glycerol 3-phosphate regulon of *Escherichia coli* K12. *J. Biol. Chem.* **1987**, *262*, 15869–15874. [PubMed]
54. Iuchi, S.; Cole, S.T.; Lin, E.C. Multiple regulatory elements for the glpA operon encoding anaerobic glycerol-3-phosphate dehydrogenase and the glpD operon encoding aerobic glycerol-3-phosphate dehydrogenase in *Escherichia coli*: Further characterization of respiratory control. *J. Bacteriol.* **1990**, *172*, 179–184. [CrossRef] [PubMed]
55. Heyland, J.; Blank, L.M.; Schmid, A. Quantification of metabolic limitations during recombinant protein production in *Escherichia coli*. *J. Biotechnol.* **2011**, *155*, 178–184. [CrossRef] [PubMed]
56. Glick, B.R. Metabolic load and heterologous gene expression. *Biotechnol. Adv.* **1995**, *13*, 247–261. [CrossRef]
57. Weissenborn, D.L.; Wittekindt, N.; Larson, T.J. Structure and regulation of the glpFK operon encoding glycerol diffusion facilitator and glycerol kinase of *Escherichia coli* K-12. *J. Biol. Chem.* **1992**, *267*, 6122–6131. [PubMed]

58. Hogema, B.M.; Arents, J.C.; Bader, R.; Postma, P.W. Autoregulation of lactose uptake through the LacY permease by enzyme IIAGlc of the PTS in *Escherichia coli* K-12. *Mol. Microbiol.* **1999**, *31*, 1825–1833. [CrossRef] [PubMed]

59. Berg, J.M.; Tymoczko, J.L.; Stryer, L. *Biochemistry*, 5th ed.; W. H. Freeman: New York, NY, USA, 2002.

60. Marr, A.G.; Ingraham, J.L.; Squires, C.L. Effect of the temperature of growth of *Escherichia coli* on the formation of beta-galactosidase. *J. Bacteriol.* **1964**, *87*, 356–362. [PubMed]

61. Mayer, S.; Junne, S.; Ukkonen, K.; Glazyrina, J.; Glauche, F.; Neubauer, P.; Vasala, A. Lactose autoinduction with enzymatic glucose release: Characterization of the cultivation system in bioreactor. *Protein Expr. Purif.* **2014**, *94*, 67–72. [CrossRef] [PubMed]

bioengineering

MDPI

Article

Hybrid Approach to State Estimation for Bioprocess Control

Rimvydas Simutis [1] and Andreas Lübbert [2,*]

[1] Department of Automation, Kaunas University of Technology, Kaunas 44249, Lithuania;
    Rimvydas.simutis@ktu.lt
[2] Department of Biochemie/Biotechnologie, Martin-Luther-Universität Halle-Wittenberg,
    06108 Halle, Germany
*   Correspondence: andreas.luebbert@biochemtech.uni-halle.de; Tel.: +49-345-5525-942

Academic Editors: Christoph Herwig and Martin Koller
Received: 26 October 2016; Accepted: 2 March 2017; Published: 8 March 2017

Abstract: An improved state estimation technique for bioprocess control applications is proposed where a hybrid version of the Unscented Kalman Filter (UKF) is employed. The underlying dynamic system model is formulated as a conventional system of ordinary differential equations based on the mass balances of the state variables biomass, substrate, and product, while the observation model, describing the less established relationship between the state variables and the measurement quantities, is formulated in a data driven way. The latter is formulated by means of a support vector regression (SVR) model. The UKF is applied to a recombinant therapeutic protein production process using *Escherichia coli* bacteria. Additionally, the state vector was extended by the specific biomass growth rate μ in order to allow for the estimation of this key variable which is crucial for the implementation of innovative control algorithms in recombinant therapeutic protein production processes. The state estimates depict a sufficiently low noise level which goes perfectly with different advanced bioprocess control applications.

Keywords: State estimation; hybrid modeling; Unscented Kalman Filter; recombinant protein production

1. Introduction

Producers of recombinant therapeutic proteins are increasingly forced to enhance the batch-to-batch reproducibility of their cultivation runs at a high level of productivity. This is not only required to simplify the downstream processing, but it also increases the productivity. Thus, the process must be kept tightly on its predefined optimized track. The important tools for guaranteeing the quality of the cultivation processes are advanced monitoring and feedback control systems. Various bioprocess monitoring and control techniques are described in literature [1–3]. They all suffer from the flaw that the online values of the important controlled process variables (biomass, substrate, and product concentrations) are difficult to measure or to estimate to a sufficient accuracy online, and in the current papers on bioprocess control systems the determination of reliable data from the process is usually insufficiently considered.

Here we follow the current development in many engineering subjects, for example, navigation (e.g., [4]), economics (e.g., [5]), tracking moving objects, and process state estimation (e.g., [5]), to increase the estimation accuracy by means of model supported estimation algorithms which combine the a priori knowledge about the process under consideration and the actual measurement data from the online measurement devices [6–8].

The most often employed techniques for state monitoring and estimation are based on Kalman Filters, which are also used in modern bioprocess engineering. The original Kalman Filter algorithm provides optimal estimates of measured and unmeasured bioprocess state variables by combining information of linear mathematical models and online measurements [1,8,9].

Importantly, Kalman Filters do not describe the process' state simply by an N-dimensional vector. Instead, the state is considered a set of N random variables which, at a given time instant, are described by means of probability distribution density functions. The propagation of these density functions with time step is computed using the dynamical process model. As the original Kalman Filter [10] assumes the model to be linear, the propagation of the density function, such as an N-dimensional Gaussian distribution, remains a Gaussian distribution at all times. Such linear propagations do not change the shape of a probability distribution. The new spread of the resulting Gaussian is determined by a new covariance matrix. In nonlinear processes this does not hold. Nonlinear propagations usually change the form of a Gaussian distribution and result in another distribution with at least a skew.

Since most real bioprocess systems are nonlinear, and the state variables are strongly coupled with each other, various extensions of the standard Kalman Filter procedure were proposed. In processes with not too strong nonlinearities, the time increments can be kept so small that the model can be linearly approximated at each time step using a first-order Taylor series linearization of the nonlinear model in order to compute the new covariance matrix. This approach is used in extended Kalman Filters (EKFs), where the density function can be propagated as in the original Kalman Filter. This means that one uses a linear model approximation in order to keep a Gaussian distribution density Gaussian.

In Unscented Kalman Filters (UKFs) one takes another way [11–13]. Here the original nonlinear models are directly used and not changed in any way. As these do not conserve the functional form of the density functions upon a mapping to the next time instant, the resulting distribution densities are corrected. As long as the multidimensional density functions are dominated by their means and their covariances, they can be characterized by a small number of points. A simple one-dimensional Gaussian, for instance, is fully determined by its mean μ and its spread (i.e., its mean μ and its inflection points on both sides of the mean at $\mu \pm \sigma$, where σ is the standard deviation). Analogously, in the Unscented Kalman Filter with N state variables, one assumes the distribution density function to be a Gaussian and takes $2N + 1$ points, the so-called sigma-points, to characterize its form. In order to move the density function from time step to time step, one moves theses characteristic points separately in the state space using the dynamic model. Then, from the transferred points one then determines the mean and the corresponding variances/covariances of the new Gaussian bell. Hence, it is assumed that the new Gaussian distribution density function can be described all the way by means and covariances.

The decisive advantage of the Unscented Kalman Filter is that the dynamic process model and the model that relates the state variables to the quantities that are measured online can be used in their original forms. Only the description of the uncertainty of the states is approximated. The computational complexity of the UKF is similar to that of the Extended Kalman Filter (EKF), and can thus be implemented into the commonly used automation systems.

Although further developments of Kalman Filters also allow for the removal of the restriction on the description of the random variables to simple density functions (an example is the particle filter), they currently cannot be recommended for state estimation in biotechnology. Their relatively small improvements in the estimation accuracy do not justify the much higher computational expenditure they require. Hence, we will consider the Unscented Kalman Filter and propose a hybrid combination of a conventional system of ordinary differential equations to compute the propagation of the state from time step to time step and a data driven model in the form of a Support Vector Machine (SVM) for the mapping of the predicted state variable to the measurement quantities.

2. Experimental Data

The data taken here to demonstrate the procedure were taken from Schaepe et al. [14]. There, an *Escherichia coli* strain (BL21:DE3 pLysS) was used which expressed the green fluorescent protein sfGFP [15] under control of the T7-promoter upon induction with IPTG. This protein becomes active within *E. coli*'s cytoplasm and can be detected within the cells with a spectro-fluorimeter.

Importantly, the specific product formation rate π increases monotonically with the cell's specific growth rate μ. The cultivations were performed in a fed-batch mode at a temperature of 30 °C and pH 7.0 in a stirred tank bioreactor with 15 L maximal working volume.

From all data produced and reported in Schaepe et al. [14], we took the records of the three validation experiments S836 to S838 to demonstrate the process supervision with the proposed hybrid version of the UKF. The corresponding feed rate profiles were determined during tracking experiments, which responded to the changing oxygen uptake capabilities of the cells.

During the cultivations, the UKF only uses the online measured offgas data signals, particularly the cumulative oxygen uptake and carbon dioxide formation rates signals, cOUR and cCPR, respectively, to estimate the biomass and product concentrations as well as the specific biomass growth rate.

These data demonstrate that the estimates which only use online measured data from the offgas analysis very closely predict the biomass and product concentration data, which as offline measured data became available much later only.

3. Process Modeling

Kalman Filters [10] require two models; the first is used to move the elements of the state vector within the state space from one time step t_{k-1} to the next one t_k. The second is the observation model that relates the state vector at time t_k to the actual measurement quantities at that time. The Kalman Filter algorithm estimates the current state of process variables, along with their uncertainties.

3.1. State Propagation Model

For the state propagation model, a basic ordinary differential equation system describing the propagation of the initial state with time is used. The conventionally used equation system involves the mass balances around the reactor for the state variables. As such, we consider the biomass, the substrate, and the product. As the process was operated in the fed-batch mode, an additional equation is required that takes into account the change of the working volume *W* with time.

$$\frac{\partial c}{\partial t} = R + \frac{F}{W}(c_F - c) \tag{1}$$

$$\frac{d\mu}{dt} = 0 \tag{2}$$

$$\frac{dW}{dt} = F \tag{3}$$

Here, *c* = [X; S; P] is the state vector with the concentrations, X of biomass, S of the substrate, and P of the product. The specific growth rate μ is taken as an additional state variable which can be estimated during the state estimation procedure. It is assumed to be practically constant and only changed by some modeling noise given by the corresponding element of diagonal covariance matrix V_{mod}. *F* is the substrate feed rate, and c_F is the concentration of the solution fed to the culture. The substrate concentration is the only nonzero element in c_F. It was 600 g/L in this concrete case.

The biochemical conversion is described by the volumetric conversion rate, *R*, which contains the specific conversion rates of the biomass, μ, the substrate, σ, and the product, π. Usually these are modeled by simple or slightly extended Monod expressions. Concretely, the following volumetric conversion rates were taken.

$$R = \left[\left\{ \mu; -\left(\frac{\mu}{Y_{xs}} + \frac{\pi}{Y_{ps}} + m_s \right); Y_{px}\mu \right\} * X \right] \tag{4}$$

As the specific biomass growth rate μ was taken as a state variable, its value is taken from the current state estimate of the Unscented Kalman Filter.

With the initial conditions c_0 for c, μ_0 for the specific biomass growth rate μ, and W_0 for W, as well as the feed rate profile $F(t)$ and the concentrations c_F in the feed, the equation can be solved. The feed rate profiles are manipulated variables and could be measured online (Figure 1).

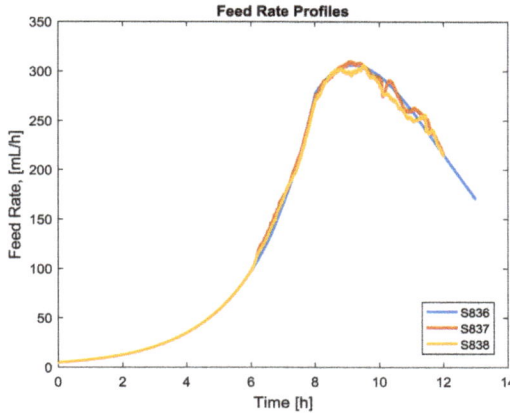

Figure 1. Feed rate profiles from the cultivation experiments used here as the example.

With the feed rate data depicted in Figure 1, the model can be fitted to process data in order to obtain the free parameters of the dynamic process model, the yields Y_{xs}, Y_{ps}, Y_{px}, and the maintenance coefficient m_s.

3.2. Observation Model

The state vector c at time t_k corresponds to a number of quantities that can be measured during the cultivation process. The obvious first question, which of the possible measurement variables reflect the most information about the process' dynamics, can quite easily be answered. For that purpose, it is straightforward to look at the well-established gross reaction equation that describes the biochemical conversion process. It contains, elementwise, the conversion of the significantly changing components with respect to the elements carbon (C), hydrogen (H), oxygen (O), and nitrogen (N). A typical equation is:

$$C_6H_{12}O_6 + a\ O_2 + b\ NH_3 = c\ CH_{1.8}O_{0.5}N_{0.2} + d\ CO_2 + e\ H_2O \tag{5}$$

It is referred to as the stoichiometric equation of the conversion process, where the coefficients a, b, c, d, and e are stoichiometric coefficients or yields.

As the equation considers only those species, the amounts of which are significantly changing during the biochemical conversion process, this equation gives a direct indication of the quantities that should be measured during the process.

Here we are led to the oxygen consumption (O_2), the base consumption (NH_3), and the carbon dioxide (CO_2) formation. The water formation cannot be considered, as its amount is negligible as compared to the water that is part of the cultivation medium. The water production rate cannot be measured accurately enough and is not considered here.

In a practical application, the corresponding rates, the oxygen uptake rate (OUR), the carbon dioxide production rate (CPR), and the base consumption rate (BCR), are usually measured online during the cultivation process. However, in order to reduce the noise level of the measurement signals, it is advisable to replace the original rate signals by their corresponding cumulative rate signals (e.g., the cumulative oxygen uptake rate cOUR). This does not only reduce the noise level, but it additionally plays to the fact that the important state variables, the biomass and the product concentrations, are cumulative quantities as well.

Hence, we are looking for models that describe the cumulative rates cOUR and cCPR as functions of the state variables c. As we usually have measurement sampling rates in the order of 1 Hz of these quantities, while the time constant of the changes in the state variables is in the order of 1 h, the cumulation does not influence the measurement information significantly. These signals follow changes in the biochemical kinetics quickly enough, a fact which was already shown in many closed loop control investigations (e.g., [16]).

The classical textbook relationships between the oxygen uptake rates and the biomass concentrations, such as variants of the Luedeking/Piret equation [17], are not accurate enough as an observation model. Hence, it is straightforward to use data driven models for this purpose, in the sense of learning from the experience with measurement data, where mechanistic models are not yet available to a comparable level of accuracy. Various forms of nonlinear regression models (polynomials, feed forward neural networks, etc.) can be used for modelling these relationships [18].

From the many possibilities, we chose the support vector machine approach [19,20], a regression technique that is an advanced kernel approach. Support vector regression (SVR) techniques require less time and expertise than the artificial neural networks to train the model. This is mainly because SVRs are trained with a structured algorithm (quadratic optimization), which has one unique solution, and it consistently produces the same results when trained with identical data and parameters. Data from new cultivation examples can easily be used to extend and improve an existing SVR model without additional tuning of the model parameters. SVR techniques are also more robust for models with multidimensional inputs.

In our Kalman Filter we need a representation of the measurement quantities cOUR and cCPR as a function of the state variables. We took these data from the recombinant protein cultivation experiments [14] and used general radial basis functions or Gaussian bells as kernels.

The observation model describing the cumulative oxygen uptake rate and the cumulative carbon dioxide production rate as nonlinear functions of the biomass concentration is presented as lines in Figure 2. The data points (symbols in Figure 2) were taken from the offline measured biomass concentrations and the cumulative OUR and CPR data measured at the corresponding time instants. As can be seen in Figure 2, all records from the three experiments were used to train the SVM model. A cross validation technique was employed using 70% of the data points for the training and 30% for a validation.

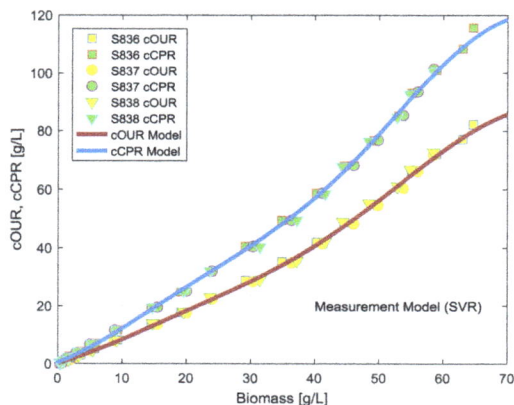

Figure 2. Cumulative oxygen uptake and carbon dioxide production rates signals as a function of the biomass concentration X. The curves show a direct evaluation of the support vector regression (SVR) model trained on the data of the cultivations S836, S837, and S838 [14] using the cross validation techniques. cOUR, cumulative oxygen uptake rate; cCPR, cumulative carbon dioxide production rate.

4. Employing the Unscented Kalman Filter

As all Kalman Filters, the Unscented Kalman Filter UKF is a recursive algorithm that determines the estimate $c(t_k)$ at time t_k from the previous estimate $c(t_{k-1})$ [11,12].

It first proposes a state vector $\hat{c}(t_k)$ from the previous estimate $c(t_{k-1})$ using the nonlinear process model Ψ, (in our concrete application, the model is described by Equation (1), where the actual state vector $c(t)$ is [X; S; P; μ]) and computes the corresponding measurement quantities $y(t_k)$ from $\hat{c}(t_k)$ using the nonlinear observation model H (in our case, this model is presented by support vector regression equations for cOUR and cCPR). The proposal $\hat{c}(t_k)$ is then corrected to compute the new estimate $c(t_k)$ using the difference between the actually measured values $y^{(m)}(t_k)$ and the computed values $y(t_k)$:

$$c(t_k) = \hat{c}(t_k) + K\,(y^{(m)}(t_k) - y(t_k)) \tag{6}$$

where the matrix K that rules the correction of the proposal $\hat{c}(t_k)$ depends on the uncertainties of the observations and the transfer model [12]. In this application, the covariance matrixes were taken as diagonal matrices. The initial state covariance matrix had the diagonal elements $V_{state} = \mathrm{diag}([.2, .2, .2, 2.0])$. The measurement noise V_{meas}, and V_{mod}, the model noise, are also incorporated with their diagonal elements $V_{meas} = [0.01, 0.01]$, and $V_{mod} = 0.01 * V_{state}$.

Figure 3 shows an example of an UKF state estimation of the biomass and the product concentration from measurement data of the cumulative oxygen uptake rate cOUR and the cumulative carbon dioxide production rate cCPR signals based on data (Cultivations S836, S837, and S838) from Schaepe et al. [14]. The symbols shown in Figure 3 are measurement data that were measured offline. They were not used during the estimate of the state variables, and are only taken to show that the estimates are accurate.

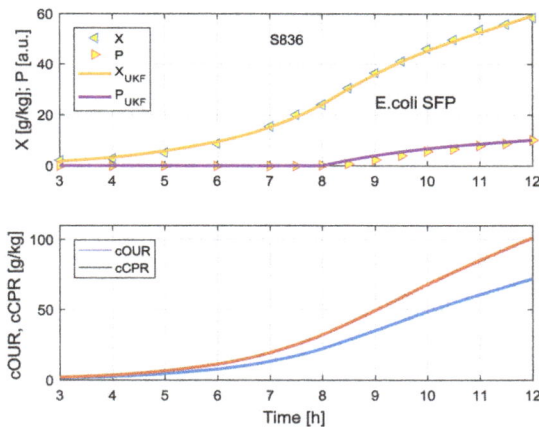

Figure 3. Typical result for the study S836: In the upper plot, the biomass and the product concentration data are displayed as symbols together with the Unscented Kalman Filter (UKF) estimates (lines). In the lower plot the measurement data used in the estimates are depicted.

The Unscented Kalman Filter software encodes the algorithm described by Wan and van der Merwe [12] (Algorithm 3.1 in that work) and was encoded in Matlab [21]. Therein the SVR regression software was used to train and evaluate the observation model. For that purpose, the generally accessible LIBSVM-software of Chang and Lin [22] was utilized and radial bases functions were used as kernel functions.

Even if the measurement values cOUR and cCPR are artificially distorted by random noise, for example, by 2.5% of the measured values, the Unscented Kalman Filter does not show much different results in the state variables biomass X and product P concentrations, as shown in Figure 4.

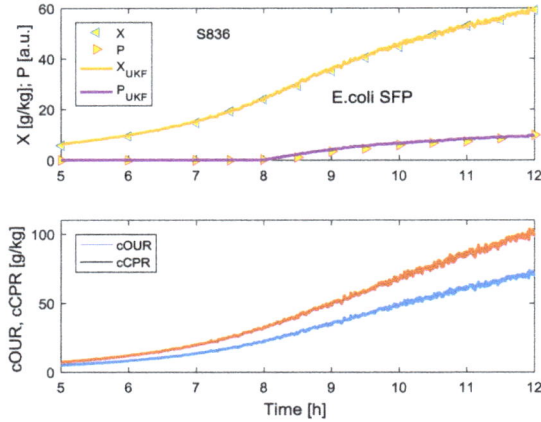

Figure 4. Results corresponding to the graphs in Figure 3 with 2.5% noise on the cOUR and cCPR measurements.

The results for the other two cultivation data records are qualitatively the same, and are thus not repeated here. As already stated above, the UKF algorithm was also used for estimating the specific growth rate of the biomass. Figure 5 presents the typical estimation result of the specific growth rate profile across the cultivation.

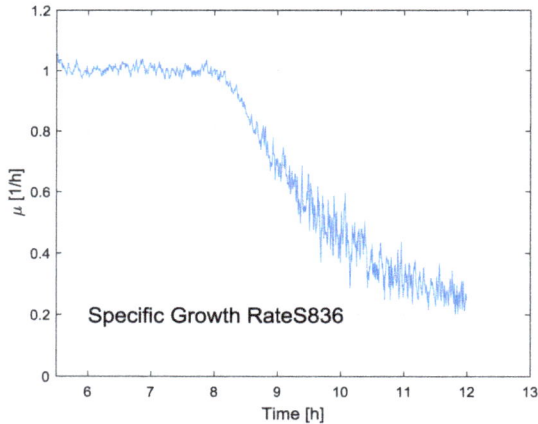

Figure 5. Estimation of the specific growth rate during the cultivation run S836.

These estimated online values of biomass, product concentrations, and specific growth rate estimates can then be used in various inferential data analysis and specific growth rate control schemes, as well as for process optimization tasks.

Bioengineering **2017**, *4*, 21

5. Conclusions

Process supervision is recommended with Unscented Kalman Filters where the dynamic equations are based on mass balances for the biomass, the substrate, and the product, and formulated by well-established ordinary differential equation systems. As the biomass growth kinetics is not a priori known on the same level of accuracy the specific biomass growth rate μ was taken as an unknown, which is estimated in the same way as the other state variables. The less well-known relationships between the state variables biomass, substrate, and product concentrations and the measurement quantities can be modelled to a sufficient degree of accuracy with modern data-driven methods developed in the machine learning community. The support vector machine technique [19] is one example, advanced neural networks and relevance vector machines [23,24] are other alternatives.

The decisive advantage of this type of nonlinear Kalman Filters is that the process and measurement models can be used directly in the estimation algorithms without any change and without the necessity of linearizing the models. The results show that the hybrid UKF method using a support vector regression model as the observation model delivers satisfactory estimates of the state variables, particularly the most important ones, the biomass and the product concentrations, and even the specific biomass growth rate. The example uses real process data in order to show that the estimation technique is not merely a play with software concepts, but leads to process data that are more accurate and reliable than the separate simulated and measured data.

These accurate estimates of the state variables are well suited for advanced process monitoring and control tasks.

Acknowledgments: The authors acknowledge the work of S. Schaepe and A. Kuprijanov who made the experiments, the data of which were used to illustrate the method presented here.

Author Contributions: Both authors jointly designed the computational method, performed the calculations and wrote the paper.

Conflicts of Interest: The authors declare no conflict of interest.

References

1. Dochain, D. *Bioprocess Control*; John Wiley & Sons: London, UK, 2008.
2. Mandenius, C.F. Recent developments in the monitoring, modeling and control of biological production systems. *Bioprocess Biosyst. Eng.* **2004**, *26*, 347–351. [CrossRef] [PubMed]
3. Simutis, R.; Lübbert, A. Bioreactor control improves bioprocess performance. *Biotechnol. J.* **2015**, *10*, 1115–1130. [CrossRef] [PubMed]
4. Gade, K. A non-singular horizontal position representation. *J. Navig.* **2010**, *63*, 395–417. [CrossRef]
5. Huang, S.C.; Wang, N.Y.; Li, T.Y.; Lee, Y.C.; Chang, L.F.; Pan, T.H. Financial forecasting by modified Kalman Filters and kernel machines. *J. Stat. Mang. Syst.* **2013**, *16*, 163–176. [CrossRef]
6. Simon, D. Kalman filtering with state constraints: A survey of linear and nonlinear algorithms. *IET Control Theory Appl.* **2010**, *4*, 1303–1318. [CrossRef]
7. Jazwinski, A.H. *Stochastic Processes and Filtering Theory*; Academic Press: Cambridge, MA, USA, 1970.
8. Simon, D. *Optimal State Estimation: Kalman, H Infinity and Nonlinear Approaches*; John Wiley and Sons: Chichester, UK, 2006.
9. Stephanopoulos, G.; San, K.Y. Studies on on-line bioreactor identification: Theory. *Biotechnol. Bioeng.* **1984**, *26*, 1176–1188. [CrossRef] [PubMed]
10. Kalman, R.E. A new approach to linear filtering and prediction problems. *J. Basic Eng.* **1960**, *82*, 35–45. [CrossRef]
11. Julier, S.J.; Ullmann, J.K. A New Extension of the Kalman Filter to Nonlinear Systems. Available online: https://people.eecs.berkeley.edu/~pabbeel/cs287-fa15/optreadings/JulierUhlmann-UKF.png (accessed on 6 March 2017).
12. Wan, E.A.; van der Merwe, R. The Unscented Kalman Filter for Nonlinear Estimation. Available online: https://www.seas.harvard.edu/courses/cs281/papers/unscented.png (accessed on 6 March 2017).

13. Gustafsson, F.; Hendeby, F.; Some, G. Some Relations between Extended and Unscented Kalman Filters. *IEEE Trans. Signal Process.* **2012**, *60*, 545–555. [CrossRef]
14. Schaepe, S.; Kuprijanov, A.; Simutis, R.; Lübbert, A. Avoiding overfeeding in high cell density fed-batch cultures of *E. coli* during the production of heterologous proteins. *J. Biotechnol.* **2014**, *192*, 146–153. [CrossRef] [PubMed]
15. Pedelacq, J.D.; Cabantous, S.; Tran, T.; Terwilliger, T.C.; Waldo, G.S. Engineering and characterization of a superfolder green fluorescent protein. *Nat. Biotechnol.* **2006**, *24*, 79–88. [CrossRef] [PubMed]
16. Gnoth, S.; Jenzsch, M.; Simutis, R.; Lübbert, A. Control of cultivation processes for recombinant protein production: A review. *Bioprocess Biosyst. Eng.* **2007**, *31*, 21–39. [CrossRef] [PubMed]
17. Luedeking, R.; Piret, E.L. A kinetic study of the lactic acid fermentation. Batch process at controlled pH. *J. Biochem. Microbiol. Technol. Eng.* **1959**, *1*, 393–412. [CrossRef]
18. Jenzsch, M.; Simutis, R.; Eisbrenner, G.; Stuckrath, I.; Lübbert, A. Estimation of biomass concentrations in fermentation processes for recombinant protein production. *Bioprocess Biosyst. Eng.* **2006**, *29*, 19–27. [CrossRef] [PubMed]
19. Vapnik, V.N. *Statistical Learning Theory*; Wiley: New York, NY, USA, 1998.
20. Smola, A.J.; Schölkopf, B. A tutorial on support vector regression. *Stat. Comput.* **2004**, *14*, 199–222. [CrossRef]
21. Mathworks. *Matlab, the Language of Technical Computing*; The Mathworks: Natick, MA, USA, 2015.
22. Chang, C.C.; Lin, C.J. LIBSVM—A Library for Support Vector Machines. Available online: http://www.csie.ntu.edu.tw/~cjlin/libsvm (accessed on 6 March 2017).
23. Tipping, M. Sparse Bayesian learning and the relevance vector machine. *J. Mach. Learn.* **2001**, *1*, 211–214.
24. Tipping, M. Bayesian inference: An introduction to principles and practice in machine learning. In *Advanced Lectures on Machine Learning*; Bousquet, O., von Luxburg, U., Rätsch, G., Eds.; Springer: Berlin, Germany, 2004; pp. 41–62.

bioengineering

MDPI

Article

Workflow for Criticality Assessment Applied in Biopharmaceutical Process Validation Stage 1

Thomas Zahel [1], Lukas Marschall [1], Sandra Abad [2], Elena Vasilieva [2], Daniel Maurer [2], Eric M. Mueller [3], Patrick Murphy [3], Thomas Natschläger [4], Cécile Brocard [2], Daniela Reinisch [2], Patrick Sagmeister [1] and Christoph Herwig [1,*]

[1] Exputec GmbH, Mariahilferstraße 147, 1150 Vienna, Austria; thomas.zahel@exputec.com (T.Z.);
    lukas.marschall@exputec.com (L.M.); patrick.sagmeister@exputec.com (P.S.)
[2] Boehringer Ingelheim RCV GmbH & Co KG, Doktor-Boehringer-Gasse 5-11, 1120 Vienna, Austria;
    sandra.abad@boehringer-ingelheim.com (S.A.); elena.vasilieva@boehringer-ingelheim.com (E.V.);
    daniel.maurer@boehringer-ingelheim.com (D.M.); cecile.brocard@boehringer-ingelheim.com (C.B.);
    daniela.reinisch@boehringer-ingelheim.com (D.R.)
[3] Versartis Inc., 4200 Bohannon Drive, Suite 250, Menlo Park, CA 94025, USA;
    emueller@versartis.com (E.M.M.); pmurphy@versartis.com (P.M.)
[4] Software Competence Center Hagenberg, Softwarepark 21, 4232 Hagenberg, Austria;
    Thomas.Natschlaeger@scch.at
*   Correspondence: christoph.herwig@exputec.com; Tel.: +43-1-58801-166400

Academic Editor: Daniel G. Bracewell
Received: 7 September 2017; Accepted: 7 October 2017; Published: 12 October 2017

Abstract: Identification of critical process parameters that impact product quality is a central task during regulatory requested process validation. Commonly, this is done via design of experiments and identification of parameters significantly impacting product quality (rejection of the null hypothesis that the effect equals 0). However, parameters which show a large uncertainty and might result in an undesirable product quality limit critical to the product, may be missed. This might occur during the evaluation of experiments since residual/un-modelled variance in the experiments is larger than expected a priori. Estimation of such a risk is the task of the presented novel retrospective power analysis permutation test. This is evaluated using a data set for two unit operations established during characterization of a biopharmaceutical process in industry. The results show that, for one unit operation, the observed variance in the experiments is much larger than expected a priori, resulting in low power levels for all non-significant parameters. Moreover, we present a workflow of how to mitigate the risk associated with overlooked parameter effects. This enables a statistically sound identification of critical process parameters. The developed workflow will substantially support industry in delivering constant product quality, reduce process variance and increase patient safety.

Keywords: retrospective power analysis; process characterization study; process validation stage 1; criticality assessment; control strategy; design of experiments

1. Introduction

Process validation of pharmaceutical processes aims to demonstrate the capability of the process to constantly deliver high product quality [1,2]. Most of the warning letters connected to process validation are raised due to flaws in stage 1 [3]. The aim of process validation stage 1 is to identify a robust process design that enables the ability to constantly deliver product quality. Therefore, it is key to identify critical process parameters (CPPs) that are likely to create risk to critical quality attributes (CQAs) and set up control strategies for these CQAs. Thereby it is possible to reduce out-of-specification (OOS) events, recalls, and ultimately risk to the patient. At process validation stage 1, it is of the highest priority not to overlook a CPP in the design of the process, which as a consequence might not be controlled properly.

In order to accomplish this goal, the following steps are commonly undertaken in industry to characterize process design following a risk-based approach:

1. Risk assessment: to identify potential influential/critical parameters for each unit operation. This is usually performed using tools such as failure mode and effect analysis (FMEA) [4,5]. Ranking of potential criticality is performed using expert knowledge, historical process data, and interdependencies identified in development data.

2. Scale down model establishment: Due to the costs related to large-scale experiments, in biopharmaceutical manufacturing it is necessary to develop appropriate scale down models (SDMs) that are appropriate to investigate the interdependency between process parameters and quality attributes.

3. Experimental designs: Design of Experiments are applied to quantify the impact of process parameters (PPs) on CQAs. Prior to conducting experiments, a priori power analysis is a good practice to evaluate if an effect that leads to a change in product quality—in the following defined as a critical effect—can be detected by the proposed design setting. Statistical power is defined as the probability that we are able to detect an effect if it is truly there [6]. This is done for a priori analysis by estimating the expected signal to noise ratio, which is thought to occur during the experiments [7]. As a result of this a priori power analysis, the number of required experiments, the intended screening range, or the design itself might be adjusted. After a sufficient power can be expected, potential influential/critical parameters are purposefully varied within experiments, which is done for each unit operation separately using the previously established SDMs.

4. Criticality assessment of process parameters by evaluating experimental designs: Identification of significant factors (rejection of the null hypothesis that the effect equals 0) at a desired significance level (typically $\alpha < 0.05$) is performed using Pareto charts and analysis of significance of regression coefficients by means of ANOVA. Misleadingly, this does not imply that for non-significant factors the null hypothesis is true and their effect is zero [8]. Rather, it indicates that the uncertainty around these factors in the range examined—often indicated by large confidence intervals around the effect—is large and critical levels cannot be excluded. Commonly, only significant factors that have been observed to impact product quality or process performance are defined as critical or key, respectively. Those which cannot be stated as significantly impacting are stated as non-critical or non-key, respectively.

5. Definition of control strategy: As a means to ensure all CQAs and quality specifications are met, a process control strategy for all critical and key process parameters must be put in place. Moreover, it has to be evaluated whether their mutual worst case setting would lead to acceptable product quality levels. Commonly for biopharmaceutical production, this is accomplished by setting normal operating ranges (NOR) and proven acceptable ranges (PAR).

Although all steps are equally important to design a robust process, we frequently observed that, in industry, steps 3, 4 and 5 are more difficult to accomplish in practice. The US Food and Drug Administration (FDA) and other agencies are not prescriptive but clearly state that statistics should be used within all stages of process validation [3]. Multiple statistical tools and software for step 3 (a priori power analysis and design of experiments) and step 4 (statistical analysis of significant parameters) exist, however, the approach of those steps as described above has two major drawbacks: (i) after making several assumptions about the expected noise in the a priori power analysis of step 4, those assumptions are not checked for validity after the experiments have been performed. Especially in biopharmaceutical engineering, reproduction and analytical variability from non-validated methods, which might be used during stage 1 of process validation, as well as unexpected non-linear effects (e.g., edge of failure experiments), may lead to increased noise in the conducted design of experiments (DoEs). (ii) Criticality and potential tightening of the NOR is only taken into account for significant parameters. This might not be sufficient since parameters with large uncertainty around the estimated effect—those effects, which might be zero, but might be very large, too—can have severe effects on product quality as well.

The first of the mentioned drawbacks can be tackled by retrospective assessment of the actually received power. Although retrospective power analysis is controversially discussed when using the observed variance and observed effect size, it is an appropriate tool when comparing the observed variance in the experiments to a pre-specified critical effect [6,9]. Frequently, retrospective power is calculated using the observed effect size, which leads to uninformative results [10].

Both issues together might lead to situations where the process shows unexpected variability during routine manufacturing. Therefore, we want to present a workflow for criticality assessment that reduces the risk to overlook critical PPs. This is demonstrated based on a process characterization study of a novel long acting human growth hormone product. Exemplarily for two unit operations, we will address the following topics:

- Establishment of a methodology that prevents engineers, during process validation, from overlooking critical parameters;
- Setting a control strategy for critical and likely overlooked parameters that ensures a robust process design;
- A workflow that can be used during stage 1 process validation to assess PP criticality. Applying those guidelines, it will be possible to better understand potential process variability and provide an opportunity to reduce process variability, OOS events, and patient risk.

2. Methods

In the following sections, we describe the biopharmaceutical production process, selection of experimental designs to study the impact of PPs on CQAs (Section 2.1), calculation procedures for critical effects (Section 2.2), an a priori power analysis approach (Section 2.3) applied to assess the ability of the DoE to detect practically relevant (here critical) effects and their statistical evaluation (Section 2.4).

2.1. Description of Process and Design of Conducted Experiments

The workflow for criticality assessment will be presented for two unit operations from a biopharmaceutical manufacturing process producing a recombinant protein. The process consists of an Escherichia coli fermentation, cell lysis, precipitation (PR), clarification (depth filtration), and three subsequent preparative chromatographic columns (CC 1/CC 2/CC 3) for purification. Finally, ultrafiltration/diafiltration is performed to adjust product concentration. For the presented case study for criticality assessment, unit operations CC 1 and the precipitation step were exemplarily chosen.

Risk assessment (FMEA) conducted by process experts showed that five and four PPs respectively, had a high risk priority number and need to be studied experimentally in respect to their influence on CQAs for CC 1 and PR, respectively (see Tables 1 and 2). Due to the number of studied PPs for both unit operations, a definitive screening design was chosen [11,12]. Except one parameter (Mixing [Yes/No] for precipitation), all DoE factors are numerically scaled. Small-scale experiments were used to conduct DoEs.

2.2. Calculation of Thresholds for Critical Effects

We formulate a critical gap (CG) as the difference between the performance at set-point conditions and the threshold for each response:

$$CG = \text{threshold}_{USL} - \bar{y}(x_{SP}) \tag{1}$$

where $\bar{y}(x_{SP})$ is the response value (here a specific concentration of an impurity) at set-point condition of manufacturing. Since we do not have lower specification limits for the studied impurities, the threshold, which must not be surpassed, is derived from the upper specification limit (USL) of drug substance (DS) specifications. The studied unit operations are at an intermediate stage of the process. We therefore, calculate the specification limit for the investigated unit operation by multiplying the final DS specifications times the mean specific clearance factors from the manufacturing scale of all unit operations in between. This approach might be refined by including knowledge on increased impurity clearance, e.g., due to spiking studies. Choosing the approach with mean specific clearances might seem conservative, however, it is desirable to reduce the risk of overestimating the specific impurity clearance. The specific clearance factors for each unit operation are defined by:

$$\text{Specific Clearance} = \text{SC} = \frac{c_{CQA,load}}{c_{CQA,pool}} \tag{2}$$

where cCQA, load and cCQA, pool are the specific concentrations (mg CQA per mg product) of the respective CQA prior to and after the unit operation.

$$\text{threshold}_{USL} = \text{USL} * \prod_{u=k}^{U} \text{SC}_u \tag{3}$$

where $u = k, \ldots, U$ is counting the unit operations from the studied kth unit operation until the last unit operation (U) which equals DS.

2.3. A Priori Power Analysis

We want to investigate if the residual error during evaluation of experimental designs (DoEs) masks effects to an extent such that they could collectively surpass a critical threshold (e.g., specification limit of a specific CQA concentration) within normal operating ranges (see Section 2.2 for calculation of thresholds). Since we are dealing with a multivariate problem, we need to identify how many parameters and to what extent each of those parameters contributes to surpassing such a critical threshold. From a sparsity assumption, it is unlikely that all effects that can be studied using a certain design (e.g., all main effects and interactions effects) are truly present. Therefore, it is a common assumption applied to many statistical packages to study only power of the total number of main effects [13].

Moreover, in multivariate analysis ($p > 1$), infinite combinations of effects of multiple parameters exist that lead to such a critical threshold being surpassed, e.g., the full effect to surpass the critical threshold might be explained solely by the first parameter (P_1) and no effect is present from the residual parameters (P_r), or a fraction of the entire effect is explained by P1 (e.g., 10%) and the residual 90% is equally explained by P_r. Overall, we are interested in the mean chance to detect any of those combinations. Per default, classical statistical software such as JMP (SAS Institute Inc., Cary, NC, USA) or DesignExpert (Stat-Ease, Inc., Minneapolis, MN, USA) only allow for fixed effect power calculation [10,13]. Here, we propose a more general method based on the assumption that the effects are randomly distributed over all parameters. Therefore, we assigned weights to the parameters and varied the fraction/weight of the entire effect that is explained by each parameter gradually between 0.0 and 1.0 (we used a step size of 0.01 in our experiments, i.e., 100 steps) and split the residual effect equally under the residual parameters: $w_i = a$, $w_{j \neq i} = (1 - a)/(p - 1)$, for $a = 0, \ldots, 1$ and $i = 1, \ldots, p$. Hence all the weights w_i sum up to 1. In total, we obtain $C = p * 100$ combinations of possible effect distributions and the resulting power values. The mean for each parameter of these recorded power values was taken as the power for this experimental design (see step 6 of the a priori workflow present below).

Herein, the following workflow for a priori power analysis can be formulated:

1. Estimate the mean ($\bar{y}_{SP}$) and variance (σ_{SP}) of the response variable from small-scale or pilot-scale experiments at set point conditions of manufacturing. We assume that residual error in the model is only due to process- and analytical variance. The latter estimate will be used to calculate the expected sum of squares of the residuals ($\widetilde{SS_{res}}$):

$$\widetilde{SS_{res}} = (n-1) * \sigma_{SP} \tag{4}$$

2. For each of the combinations (c) described above, we calculate critical effects for each parameter using its weight $w_i^{(c)}$:

$$\beta^{(c)}{}_{crit,i} = \frac{w_i^{(c)} * CG}{\max(\text{NORU}_i - sp_i, sp_i - \text{NORL}_i)} \tag{5}$$

In order to estimate the individual coefficient for the i-th parameters, from a risk-based approach, we divide by the longest distance from the set-point (sp_i) to the nearest NOR border: where NORU_i is the upper boundary of the NOR and NORL_i is the lower boundary of the NOR of the parameter i. Note that this works for a symmetric as well as asymmetric NOR.

3. Using the design matrix X, obtained for a specific experimental design, we can simulate possible $\tilde{y}$ values at the screening range using:

$$\tilde{y}^{(c)} = X\beta_{crit}^{(c)} \tag{6}$$

4. From that, the total sum of squares can be estimated:

$$\widetilde{SS_{tot}}^{(c)} = \sum_i^n \left(\widetilde{y_i y}^{(c)} - mean\left(\tilde{y}^{(c)}\right) \right) \tag{7}$$

Together with the sum of squares of the residuals, the expected coefficient of variance can be calculated:

$$\tilde{R}^{2(c)} = 1 - \frac{\widetilde{SS_{res}}}{\widetilde{SS_{tot}}^{(c)}} \tag{8}$$

5. Using Cohen's effect size (f), the non-centrality parameter λ and the critical F value (F_{crit}), the a priori power for the combination c of effects that no parameter has been overlooked can be calculated [7]:

$$f^{2(c)} = \frac{\tilde{R}^{2(c)}}{1 - \tilde{R}^{2(c)}} \tag{9}$$

$$\lambda^{(c)} = f^{2(c)} * v \tag{10}$$

6. Confidence intervals for the a priori power for the combination c were calculated according to

$$\lambda_{upp}^{(c)} = \lambda^{(c)} * c_{crit}(1 - \alpha | v)/v \tag{11}$$

$$\lambda_{low}^{(c)} = \lambda^{(c)} * c_{crit}(\alpha | v)/v \tag{12}$$

where $c_{crit}(\alpha|v)$ is the $100 * \alpha$ percentile from a χ^2 distribution with v degrees of freedom.

$$F_{crit} = F_{inv}(1 - \alpha | u, v) \tag{13}$$

7.

$$power_{apriori}^{(c)} = 1 - F_{nc}\left(F_{crit}\Big|, u, v, \lambda^{(c)}\right) \tag{14}$$

where F_{nc} is the non-central F distribution with $u = p$ (number of DoE parameters) and $v = n - u - 1$, where n is the number of observations in the DoE.

8. The mean power over all combinations of effects was estimated as the arithmetic mean of all $power_{apriori}^{(c)}$:

$$power_{apriori} = \frac{\sum_{c=1}^{C} power_{apriori}^{(c)}}{C} \tag{15}$$

2.4. Evaluation of DoEs

Multiple linear models were used to identify the relationship of the studied PPs (DoE factors, X) on the response variable (y), representing a CQA or KPI of the process, up to a residual error (ε):

$$y = \beta_0 + X\beta + \varepsilon \tag{16}$$

where X is a ($n \times p$) dimensional design matrix for n DoE runs and p DoE factors which are studied, β_0 is the intercept, β are the true effects of the DoE factors, and ε is the residual, un-modelled error vector. The un-modelled error vector describes the analytical and process variance as well as non-linear effects which cannot be accounted for in the model structure. Identification of significant parameters was done using stepwise regression within the multiple linear regression (MLR) tool of inCyght software (inCyght version 2017.03, Exputec GmbH). Parameters showing a partial p-value below 0.05 were allowed to enter the model. Those which showed a p-value larger than 0.1 were excluded from the model. Starting with the most significant parameter, this including/excluding procedure was applied iteratively and was repeated till the model structure did not change any more and the optimal model was achieved by this approach; identified significant parameters and their respective p-value are shown in Tables 1 and 2 for CC 1 and PR, respectively. The normalized raw data are given in the Supporting Information Tables S1 and S2.

Table 1. p-values of significant process parameters that were used in the statistical models for each critical quality attributes (CQA) of CC 1. Normal operating ranges and thresholds are given for each process parameter or critical quality attribute, respectively. Non-significant parameters are indicated with "-". Also, the ratio of standard deviation of raw residuals of the model by the standard deviation at set-point ($\frac{\hat{\sigma}_{residues}}{\hat{\sigma}_{SP}}$) is given for each CQA.

		End Pooling [CV]	Elution Strength [mM]	Wash Strength [mM]	Column Loading Density [g/L]	pH [–]	$\frac{\hat{\sigma}_{residues}}{\hat{\sigma}_{SP}}$
CQA	NOR [1]	−1.1–0	−1.1–0.65	−1.1–1.1	−0.51–1.1	−0.55–0.55	
	Threshold						
Process impurity 2 clearance	0.85	-	-	0.059	0.099	-	7.79
Product impurity 1 clearance	1.08	0.028	-	0.098	0.089	0.027	18.12
Product impurity 2 clearance	0.1	-	-	-	-	-	256.06

[1] NOR was normalized by the screening range.

Table 2. *p*-values of significant process parameters that were used in the statistical models for each CQA of precipitation (PR). Normal operating ranges or thresholds are given for each process parameter or critical quality attribute. Non-significant parameters are indicated with "-". Also, the ratio of standard deviation of raw residuals of the model by the standard deviation at set-point ($\frac{\hat{\sigma}_{residues}}{\hat{\sigma}_{SP}}$ is given for each CQA.

		Temperature [°C]	Time [Hours]	Mixing [Yes/No]	pH [–]	$\frac{\hat{\sigma}_{residues}}{\hat{\sigma}_{SP}}$
CQA	NOR [1]	−1.71–0.41	0.33–0.41	−0.95–0.95	−0.61–0.61	
	Threshold					
Process impurity 1 concentration specific	9×10^5	9×10^{-5} *	-	-	0.07	64.89
Process impurity 2 concentration specific (prior filtration)	9×10^4	-	-	-	-	2.68
Process impurity 2 concentration specific (post filtration)	784.7	-	-	-	0.021	0.55

[1] NOR was normalized by the screening range. * A quadratic effect was modelled for temperature and the shown *p*-value corresponds to the quadratic effect.

3. Results and Discussion

Experiments performed in biotechnological studies might contain data that violate the statistical assumptions of parametric tests (i.e., normality, homogeneity of variances and independence of errors). Moreover, with a limited number of experiments and a large number of unknown parameters, such assumptions are hard to assess. Consequently, nonparametric approaches bear potential and we want to present a novel permutation test to assess the power of individual DoE factors in a multivariate regression model.

3.1. Permutation Test for Retrospective Power Analysis

The following permutation approach is adapted from a permutation test aiming to investigate power retrospectively [14]. Here, we adapted this approach to study the significance of the alternative hypothesis that critical effects are present. The following steps are performed:

1. Using variable selection procedures, we select a significant regression model (all included effects are not 0 to a certain significance level):

$$y = \beta_0 + \beta_s * X_s + R_{y|X_s} \qquad (17)$$

where X_s denotes the s significant parameters selected from a variable selection procedure (e.g., stepwise variable selection) and $R_{y|X_s}$ are the residuals of the obtained model. A list of those significantly selected parameters for the case studies of this work can be found in Tables 1 and 2.

2. We define a critical gap (CG) that we must not surpass as the difference of the threshold and the worst case model prediction within the NOR ($x_{worst\ case_{NOR}}$), which is the parameter setting where the model prediction ($\hat{y}(x)$) is closest to the threhsold$_{USL}$:

$$CG = theshold_{USL} - \hat{y}(x_{worst\ case_{NOR}}) \qquad (18)$$

3. Similar to the approach discussed in Section 2.3 for the a priori power analysis, for non-significant parameters, a variety of combinations (in total C) of effects for those parameters exist that lead to surpassing a critical threshold. In order to estimate the mean likelihood of not overlooking a specific parameter, we vary the relative impact on the threshold of each parameter gradually between 0 and 1 in 100 steps. The fraction of the CG which is attributed to the non-significant

parameter i is expressed as the weight $w_i^{(c)}$ for the combination c. Equation (5) can be used to calculate the critical effect of the parameter i.

4. The residuals $R_{y|X_s}$ are permuted randomly, producing $R^*{}_{y|X_s}$.

5. New response values are calculated from the permuted residuals assuming that the critical effect is present under the alternative hypothesis (H_A):

$$y^* = \beta_0 + \beta_s * X_s + \beta^{(c)}{}_{crit} * Z + R^*{}_{y|X_s} \qquad (19)$$

where $\beta^{(c)}{}_{crit}$ is a vector of regression coefficients for the non-significant parameters and Z is the design matrix for all non-significant parameters.

6. Make a model for y^* based on X and Z and record significance of $\hat{\beta}_{crit}$ at a certain significance level (here $\alpha = 0.05$)

7. Repeat steps 4, 5 and 6 a large number of times (here 1000) and count the number of significant outcomes for each $\hat{\beta}_{crit,i}$ at a certain significance level (here $\alpha = 0.05$). The fraction of significant outcomes of all iteration cycles equals the retrospective power of parameter i.

3.2. Comparison of a Priori and Retrospective Power

If we apply the proposed retrospective power analysis permutation test of Section 3.1 to experimental data recorded from two unit operations (CC 1 and PR), we obtain power values for each PP/CQA combination from Tables 1 and 2, respectively.

Figure 1A shows a comparison of the retrospective and a priori power analysis for the CC 1 unit operation. For all three studied CQAs at this stage ('process impurity 2 clearance', 'product impurity 2 clearance' and 'product impurity 1 clearance'), we obtain a priori estimates of 1 (rightmost bar group in Figure 1A). This indicates an ideal case to start with experiments since there is no chance of overlooking a critical effect. Retrospective power analysis revealed that all investigated PPs power values are well below the common statistical practice cut-off value of 0.8. This can be explained by the fact that the residual variance in the model is much higher than the initial estimate at the set point, expressed by ratios of $\frac{\hat{\sigma}_{residues}}{\hat{\sigma}_{SP}}$ well above 1, as shown in Table 1. In general, multiple reasons for this discrepancy between the initial guess of expected variance and the actual residual variance in the model might exist. It could be a non-representative selection of set-point runs (e.g., runs conducted with different operators), unexpected increase of variance during experiments (e.g., it is more difficult to control experiments at unusual parameter settings) or even non-linear dependency which cannot be captured by the linear model structure. Although statistically good practice, our experience shows that such non-linear dependencies might not be obvious from analysis of residuals (e.g., investigation of plots of residual vs. DoE factors). In a DoE approach, each experiment is unique in its settings if we do not use replicates and thereby no redundancy is available to hinder the model from being leveraged by non-linear responses.

For the precipitation step (PR), a priori power analysis again suggested a power of 1 (Figure 1B). Retrospectively assessed power values match the results obtained from a priori analysis, indicating that the performed DoE had sufficient power to assess critical effects of process parameters on quality attributes. This is reasonable since ratios of $\frac{\hat{\sigma}_{residues}}{\hat{\sigma}_{SP}}$ are closer to 1 for this unit operation compared to CC 1, as shown in Table 1.

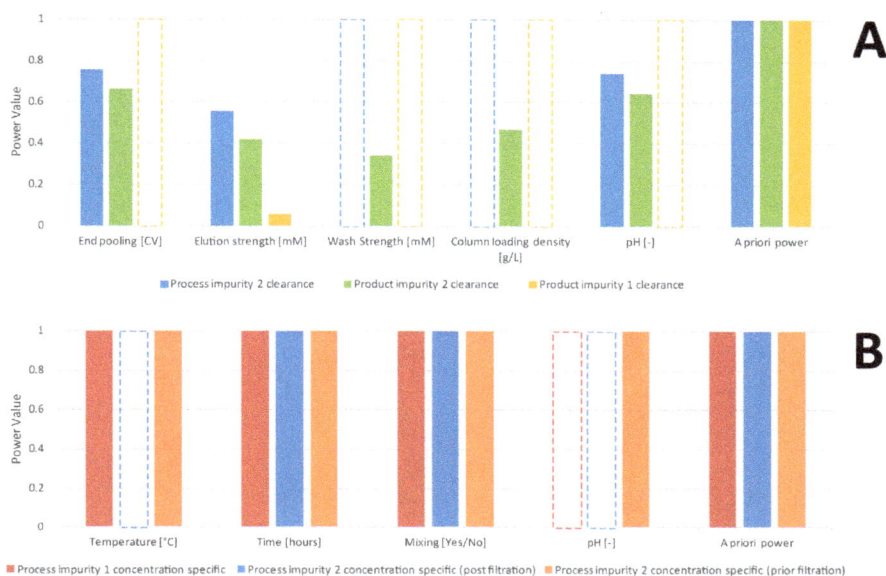

Figure 1. Power values for chromatographic column (CC) 1 (**A**) and PR (**B**) for each process parameter (PP) and CQA. Where significant process parameters were detected for a quality attribute, bars are marked grey. (**A**) Though a priori power analysis suggested a power of 100% for each investigated CQA for chromatography step 1, retrospective power analysis revealed that the power to detect a critical effect did not surpass 80% for any of the investigated process parameters. Strategies to tackle these low-power-situations are given in Figure 4. (**B**) For the precipitation step, a priori power analysis suggested a power of 100% for each investigated CQA as well. Retrospective power confirmed the findings that there is a 100% chance that we did not overlook a critical effect of the investigated process parameters on quality attributes.

3.3. How to Deal with Low-Powered Parameters?

The most common approaches to tackle insufficient power values in screening designs are by increasing the sample size, reduction of measurement variance (either analytical or process), increasing the screening range if technically possible, or accepting the lack of power, however stating the parameter as key or critical. The latter strategy will have an impact on the extended monitoring of such parameters during a subsequent process performance qualification (PPQ) campaign and routine manufacturing. As seen in Section 3.2, a priori power analysis suggested high power values for all investigated unit operations, however, drastically overestimated the power for CC 1. In specific cases, retrospectively increasing the sample size or the screening range might not be possible due to shortage of starting material or technical limitations. A measurement method with less variance might not be at hand to re-measure backup samples. Another approach made possible by the presented method for power analysis is to narrow the NOR of some process parameters. If the threshold stays the same and the NOR is symmetrically located around the set point, for smaller NORs larger effect sizes are necessary to surpass the critical threshold as shown in Equation (5) (i.e., steeper slopes). As a demonstrating scenario, we have chosen the relatively low power for Product impurity 2 clearance on CC 1 (see Figure 1A). For this response, no significant parameter could be found. Figure 2 shows how a reduction of the NOR of the process parameter, 'wash strength', impacts the power of all studied PPs of this unit operation. Upon reducing the initially defined NOR by 50% of its width, the power value for 'wash strength' increases from 0.34 to 0.68. As seen in Figure 2, the power values of the residual process parameters' effects on the same quality attribute remained unaffected, neglecting the residual variation caused by the Monte Carlo approach in permutation.

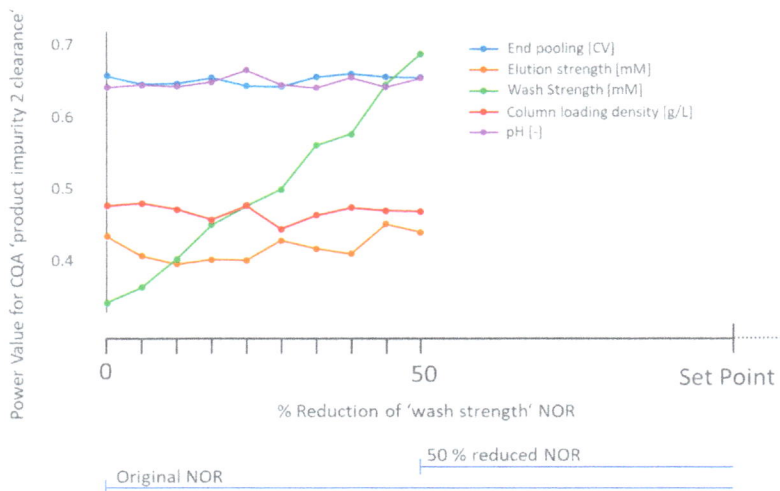

Figure 2. Retrospective power values for 'product impurity 2 clearance' for unit operation CC 1 as a function of tightened NOR of process parameter 'wash strength'. At the initially defined NOR, the power value is 0.34. Upon reducing the NOR symmetrically by 50%, the power value for this process parameter increases to 0.68. The power values of the residual process parameters remain unaffected. The visible variation can be attributed to the variance in the permutation test.

This provides an opportunity to implement a tighter control strategy though adjusting the NOR as an approach to ensure no critical effects have been overlooked. However, it may not be technically feasible or desirable for all process parameters to implement a tighter control strategy with narrower ranges, especially for a parameter that has not been confirmed to significantly impact a CQA. Since a process parameter is studied in respect to multiple CQAs, we want to note that the tightening of a NOR of a process parameter that significantly impacts one specific CQA will also increase the capacity to not overlook this parameter regarding all other CQAs which have been studied in the same experiment. In contrast to changing the NOR of a non-significant parameter onto a CQA as shown for the combination 'wash strength' onto 'product impurity 2 clearance' in Figure 2, we investigated how the change of a significant parameter impacts power levels (Figure 3). This was exemplarily done for a decrease in NOR of 'wash strength' and we recorded power values for 'process impurity 2 clearance' of all non-significant parameters as (here End pooling, elution strength and pH), as shown in Figure 3. We can see that due to the reduction of the NOR of a significant parameter, the power values of all non-significant parameters increase too. In detail, a 50% reduction of the NOR of the significantly impacting parameter 'wash strength' increases the power of all non-significant parameters by approximately 10%. This can be explained by the fact that the worst case model prediction within the reduced NOR leads to a larger CG as defined in Equation (18). Thereby, the critical effects will also be larger (Equation (6)) and consequently the chances of overlooking larger critical effects will be reduced. In this way, an improved control strategy for a known significant parameter would improve the confidence that all residual non-significant parameters were not overlooked. This is potently a more desirable approach as improved control of known significant parameters is typically required and advantageous, if feasible.

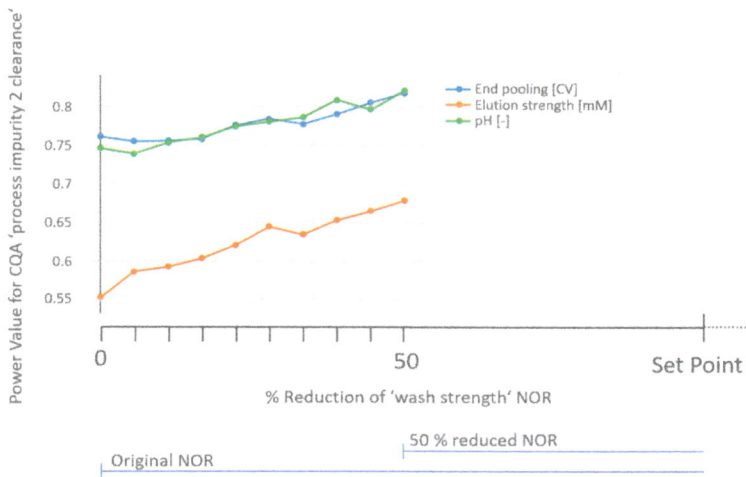

Figure 3. Retrospective power values for 'process impurity 2 clearance' for unit operation CC 1 as a function of tightened NOR of process parameter 'wash strength'. Since wash strength and column loading density are significant parameters in this model, the power was not assessed for those two parameters. Upon reducing the NOR symmetrically by 50% of the significant parameter 'wash strength', power values of all other parameters increase since the critical gap is increased, too, due to a reduction of the worst case model prediction in the NOR (Equation (18)).

3.4. Workflow for Criticality Assessment

In order to summarize the knowledge obtained from the application of the proposed posterior power analysis on two unit operations, we present a workflow that should aid process engineers in assessment of critical parameters (Figure 4). After selection of design and appropriate experiment number, a priori power analysis identifies if it is likely that a critical effect will not be overlooked. Sufficient power levels are normally assumed at 0.8 to 0.9. In cases where sufficient power cannot be assumed, the number of experiments, type of design or screening range must be increased. Both add to the expected signal to noise ratio. When increasing the screening range, care must be taken not to incur failure in experiments due to technical limitations or likely interaction effects (edge of failure). In order to reduce the risk of edge of failure experiments, it is beneficial to conduct an expected worst case scenario of the process parameters first and potentially revise the screening range afterwards.

In case sufficient power can be assumed, experiments can be conducted and regression modelling can be performed together with selection of significant DoE factors/parameters. After the "optimal" model was selected with its significant factors, retrospective power analysis, as shown in Section 3.1, will estimate the chances that the residual non-significant factors might contribute to effects that surpass a pre-specified critical threshold. In case all non-significant parameters show power values well above 0.8 to 0.9, all of them can be stated as non-critical since the residual chance that they have been overlooked is only 20 to 10%, respectively. Otherwise, for those parameters that show insufficient power, analytical and/or reproducibility variance might be lowered by re-measurement of the samples or re-conducting of experiments, respectively. Another option is to narrow the NOR of potentially overlooked parameters which show large variability. This decreases their respective critical effect according to Equation (5). After one of those three countermeasures has been taken, retrospective power analysis can be repeated to ensure sufficient power values are reached and all parameters can be stated as non-critical. If none of the above three options is technically feasible or desirable, potentially overlooked parameters should be stated as critical and monitored during process performance qualification (PPQ) runs or routine manufacturing.

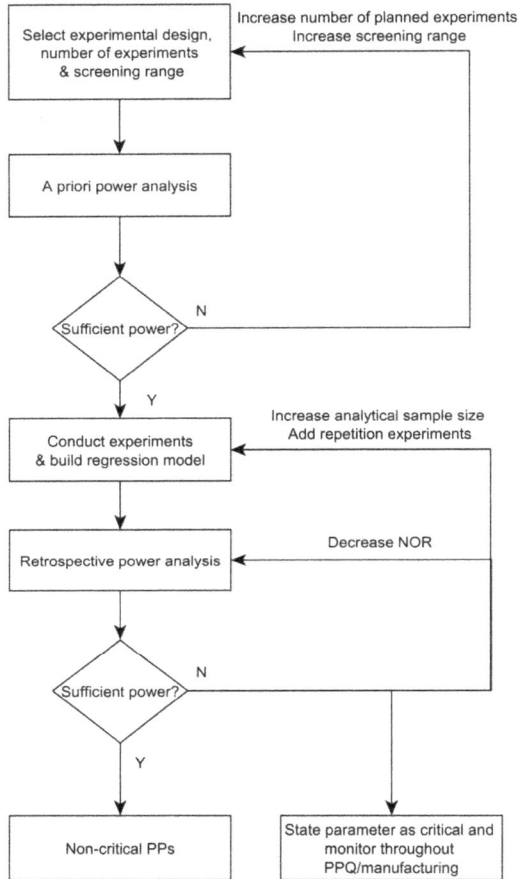

Figure 4. Workflow for criticality assessment of process parameters during process validation stage 1.

4. Conclusions

The goal of the contribution was to demonstrate the capability of a multivariate retrospective power analysis methodology to identify critical process parameters during pharmaceutical process validation stage 1.

We have shown in a case study that parameters that are non-significant in models, which were initially thought to be sufficiently powerful to identify critical effects, might still show effects that surpass a critical threshold due to increased analytical, process, or reproducibility variance. This leads to situations where the impact of those parameters on final drug product quality cannot be excluded. This was shown using a biopharmaceutical case study conducted at a world leading CMO. However, common practice is to state such parameters as non-critical and thereby overlook their potential harmful impact. Therefore, two missing parts have been introduced in this contribution: (i) a novel permutation methodology for multiple linear regression that estimates retrospective power (i.e., the chance of non-significant parameters to mutually combine to a critical effect) and (ii) a workflow for

Bioengineering **2017**, *4*, 85

criticality assessment that shows strategies of how to mitigate the risk of low-powered parameters. Besides the well-known fact that an increase in experiments increases power, it could be shown that a reduction of the NOR of significant parameters increases the power of all non-significant parameters via a reduction of the worst case model predictions; a reduction of the NOR of a specific non-significant parameter increases power solely for this parameter. Additionally, if implementation of tighter NOR controls is practically infeasible, this methodology can, at a minimum, appropriately assess the process risk and increase awareness of the limitations of the initial classification, potentially suggesting that an improved control strategy is required.

Using both tools, it will be possible for process engineers during the design stage of a process validation (stage 1) to:

- reduce the chance of overlooking potential CPPs
- develop a control strategy for potentially overlooked CPPs in order to increase process robustness
- lower OOS events and finally contribute to increased patient safety.

Supplementary Materials: The following are available online at www.mdpi.com/2306-5354/4/4/85/s1, Table S1: Standardized experimental data from DoE study of primary recovery (PR), as well as upper and lower normal operating ranges (NOR_U, NOR_L, respectively) and scale down model (SDM) variance and mean. Normalization was performed by subtracting all values by the mean and diving by the standard deviation of DoE runs, Table S2: Standardized experimental data from DoE study of chromatography column 1 (CC1), as well as upper and lower normal operating ranges (NOR_U, NOR_L, respectively) and scale down model (SDM) variance and mean. Normalization was performed by subtracting all values by the mean and diving by the standard deviation of DoE runs.

Author Contributions: Thomas Zahel developed the retrospective power analysis and criticality assessment workflow and wrote the manuscript. Lukas Marschall assisted in the development of the retrospective power analysis method, wrote the manuscript and designed the illustrations. Thomas Natschläger assisted in the development of the retrospective power analysis method and review of the manuscript. Eric M. Mueller, Pat Murphy, Sandra Abad, Cécile Brocard, Daniela Reinisch, Patrick Sagmeister and Christoph Herwig assisted in writing and reviewing the manuscript. Sandra Abad, Elena Vasilieva and Daniel Maurer conducted the necessary experiments in DoE approaches at small scale.

Conflicts of Interest: The authors declare no conflict of interest.

References

1. Ahir, K.B.; Singh, K.D.; Yadav, S.P.; Patel, H.S.; Poyahari, C.B. Overview of Validation and Basic Concepts of Process Validation. *Sch. Acad. J. Pharm.* **2014**, *3*, 178–190.
2. FDA Guidance for Industry: Process Validation: General Principles and Practices. 2011. Available online: https://www.fda.gov/downloads/drugs/guidances/ucm070336.pdf (accessed on 10 October 2017).
3. Katz, P.; Campbell, C. FDA 2011 process validation guidance: Process validation revisited. *J. GXP Compliance* **2012**, *16*, 18–29.
4. Mollah, A.H. Application of failure mode and effect analysis (FMEA) for process risk assessment. *BioProcess Int.* **2005**, *3*, 12–20.
5. ICH Harmonised Tripartite Guideline. Pharmaceutical Development Q8 (R2). Current Step 4 Version. Available online: https://www.ich.org/fileadmin/Public_Web_Site/ICH_Products/Guidelines/Quality/Q8_R1/Step4/Q8_R2_Guideline.pdf (accessed on 10 October 2017).
6. Peres-Neto, P.R.; Olden, J.D. Assessing the robustness of randomization tests: Examples from behavioural studies. *Anim. Behav.* **2001**, *61*, 79–86. [CrossRef] [PubMed]
7. Cohen, J. *Statistical Power Analysis for the Behavioral Sciences*; Revised Edition; Academic Press: New York, NY, USA, 1977; ISBN 978-0-12-179060-8.
8. Nickerson, R.S. Null hypothesis significance testing: A review of an old and continuing controversy. *Psychol. Methods* **2000**, *5*, 241–301. [CrossRef] [PubMed]
9. Thomas, L. Retrospective Power Analysis. *Conserv. Biol.* **1997**, *11*, 276–280. [CrossRef]
10. Thomas, L.; Krebs, C.J. A review of statistical power analysis software. *Bull. Ecol. Soc. Am.* **1997**, *78*, 126–138.
11. Jones, B.; Nachtsheim, C.J. A class of three-level designs for definitive screening in the presence of second-order effects. *J. Qual. Technol.* **2011**, *43*, 1–15.

12. Tai, M.; Ly, A.; Leung, I.; Nayar, G. Efficient high-throughput biological process characterization: Definitive screening design with the Ambr250 bioreactor system. *Biotechnol. Prog.* **2015**, *31*, 1388–1395. [CrossRef] [PubMed]
13. Shari, K.; Pat, W.; Mark, A. *Handbook for Experimenters*; Stat-Ease, Inc.: Minneapolis, MN, USA, 2005.
14. Freedman, D.; Lane, D. A Nonstochastic Interpretation of Reported Significance Levels. *J. Bus. Econ. Stat.* **1983**, *1*, 292–298.

bioengineering

MDPI

Article

Integrated Process Modeling—A Process Validation Life Cycle Companion

Thomas Zahel [1], Stefan Hauer [1], Eric M. Mueller [2], Patrick Murphy [2], Sandra Abad [3], Elena Vasilieva [3], Daniel Maurer [3], Cécile Brocard [3], Daniela Reinisch [3], Patrick Sagmeister [1] and Christoph Herwig [1,*]

[1] Exputec GmbH, Mariahilferstraße 147, 1150 Vienna, Austria; thomas.zahel@exputec.com (T.Z.); stefan.f.hauer@gmx.at (S.H.); patrick.sagmeister@exputec.com (P.S.)
[2] Versartis Inc., 4200 Bohannon Drive, Suite 250, Menlo Park, CA 94025, USA; emueller@versartis.com (E.M.M.); pmurphy@versartis.com (P.M.)
[3] Boehringer Ingelheim RCV GmbH & Co KG, Doktor-Boehringer-Gasse 5-11, 1120 Vienna, Austria; sandra.abad@boehringer-ingelheim.com (S.A.); elena.vasilieva@boehringer-ingelheim.com (E.V.); daniel.maurer@boehringer-ingelheim.com (D.M.); cecile.brocard@boehringer-ingelheim.com (C.B.); daniela.reinisch@boehringer-ingelheim.com (D.R.)
* Correspondence: christoph.herwig@exputec.com; Tel.: +43-1-997-2849

Academic Editor: Liang Luo
Received: 9 September 2017; Accepted: 12 October 2017; Published: 17 October 2017

Abstract: During the regulatory requested process validation of pharmaceutical manufacturing processes, companies aim to identify, control, and continuously monitor process variation and its impact on critical quality attributes (CQAs) of the final product. It is difficult to directly connect the impact of single process parameters (PPs) to final product CQAs, especially in biopharmaceutical process development and production, where multiple unit operations are stacked together and interact with each other. Therefore, we want to present the application of Monte Carlo (MC) simulation using an integrated process model (IPM) that enables estimation of process capability even in early stages of process validation. Once the IPM is established, its capability in risk and criticality assessment is furthermore demonstrated. IPMs can be used to enable holistic production control strategies that take interactions of process parameters of multiple unit operations into account. Moreover, IPMs can be trained with development data, refined with qualification runs, and maintained with routine manufacturing data which underlines the lifecycle concept. These applications will be shown by means of a process characterization study recently conducted at a world-leading contract manufacturing organization (CMO). The new IPM methodology therefore allows anticipation of out of specification (OOS) events, identify critical process parameters, and take risk-based decisions on counteractions that increase process robustness and decrease the likelihood of OOS events.

Keywords: process validation; process characterization study; holistic process model; predict out of specification events; Monte Carlo simulation; biopharmaceutical manufacturing

1. Introduction

The main goal of pharmaceutical manufacturing is to constantly deliver high product quality, which is reflected in regulatory guidelines [1–3]. Process validation is a major initiative to demonstrate the capability of meeting this goal and is separated in three stages (stage 1 to 3). Stage 1 aims at establishing a process design in which process variation in critical quality attributes (CQAs) is understood and connected to critical process parameters. This is usually done within a process characterization study using design of experiment (DoE) strategies. Resulting critical process parameters that have an effect on product quality require sufficient control strategies. Stage 2 consists of process performance qualification (PPQ) runs to confirm the design of the process and ensure it can consistently deliver high product quality. Stage 3, continued process verification (CPV), is an ongoing evaluation and monitoring to confirm the process remains in a state of control or to identify if new interdependencies between process parameters (PPs) and CQAs arise. Those three stages can be seen interlinked to each other as a lifecycle, where potential changes and associated risk in PPQ or routine manufacturing must be iteratively evaluated together with knowledge gained from initial process design [4]. Insufficient risk estimation of the entire process at stage 1 of process design (e.g., in terms of estimation of out of specification events) can lead to inconsistent or unpredicted performance at later stages.

Risk evaluation of individual unit operation of a pharmaceutical processes is commonly conducted by following steps in accordance with ICH Q8 guideline [2]:

- Risk assessment using knowledge of process experts, which leads to a candidate set of potential critical PPs for each unit operation.
- Experimental investigation of the impact of potentially critical PPs onto CQAs. This is usually performed in DoE approaches and statistical regression modeling is used to describe the relationship between significantly impacting critical PPs and CQAs mathematically.
- Comparison of the output of statistical model predictions within normal operating ranges or a design space to pre-defined acceptance limits for each unit operation.
- The risk of not meeting acceptance limits is mitigated by applying an appropriate control strategy, such as a reduction of the normal operating range.

One difficulty, especially in biopharmaceutical manufacturing where multiple unit operations are stacked together and critical PPs interact, is an appropriate evaluation of risk related to impurities. Risk analysis is impeded since propagation of impurities is rarely assessed holistically but rather evaluated on each unit operation separately [5]. Impurity propagation through multiple unit operations is difficult to study with reasonable representative experimental effort, especially at early stages of process design where only a limited number of manufacturing runs is available. However, simulations and modeling are necessary and useful to assess the chance of out of specification events. Having such a predictive tool in place to develop robust processes by incorporating knowledge acquired during process development and characterization experiments, unexplained variance in product quality possibly leading to recalls, complaints, and patient risk can be reduced. Therefore, it is desirable to formulate holistic process and production control strategies that prevent out of specification (OOS) events which could have already been anticipated within the design phase [6]. However, to the best of our knowledge, it has not been shown so far how a holistic risk evaluation spanning over multiple unit operation can be performed at process validation stage 1 and used to demonstrate overall process capability.

MC simulation is a tool to incorporate random variability into the modeled system and connect single modeling-units together. A random sampling distribution for the model parameters (inputs) needs be defined a priori, which does not need to be necessarily normally distributed. Within each cycle of the MC simulation, a different random set of inputs is drawn leading to discrete model results (outputs). Since a large number of MC cycles are performed, it is possible to aggregate the discrete model outputs to a predictive distribution of those outputs. Using this distributional information, it is possible to calculate probabilities of events (e.g., OOS). MC simulations have shown great potential in pharmaceutical industry for drug discovery and simulation of clinical trials [7] and is also routinely utilized for error propagation [8]. However, it has, to our knowledge, not been applied to impurity propagation of a batch-wise pharmaceutical processes.

Herein, we describe the development of an integrated process model (IPM) that is capable of capturing development and design data from multiple unit operations and is able to predict the risk of OOS probabilities through Monte Carlo simulation even at the early design stage of process validation. Moreover, we identify how variance and changes in set point of process parameters impacts drug substance quality. The model can be enriched at later stages also with data from PPQ, routine manufacturing, or additional development. Thereby, a continuous process data management is enabled and risk-based decision making during change and deviation management in continuous manufacturing can be based on the full spectrum of development, design, and manufacturing data.

At this stage, the following derived acceptance criteria for the IPM can be formulated:

- Prove process robustness of an existing design space: Prove that under normal manufacturing conditions it is unlikely to miss drug substance specification for defined CQAs
- Test process robustness under accelerated variance of process parameters and increased impurity burden
- Establish a platform that leverages process knowledge from PV stage 1 for further usage within PPQ and CPV (Stage 2 and 3 of process validation)

With this contribution we present the development of an IPM, validate the IPM using large scale manufacturing data, and demonstrate the capability of the IPM in estimating OOS probabilities under normal and accelerated conditions. This case study was recently conducted at a leading biopharmaceutical CMO in contract development of a novel long acting human growth hormone product.

2. Materials and Methods

Here we want to summarize the required inputs for the IPM as well as their assumptions that must be met in order to ensure reliable prediction of the IPM (for details see referred sections):

- Description of the process, order of unit operations, and variance of PPs under normal operating conditions (see Section 2.1). It is assumed that estimation of variance of PPs is representative for routine manufacturing.
- Optional: If initial unit operation of the process is not modeled by the IPM the starting distribution of each CQA needs to be estimated at the starting unit operation of the IPM. It is assumed that the estimation of starting distribution is representative for the real CQA distribution under routine manufacturing (see Section 2.2).
- Statistical regression models that describe significant relationships between PPs and CQAs for each unit operation (see Section 2.3.1). It is assumed that scientifically sound analytical methods (high accuracy, precision, robustness, selectivity, etc.) have been used to record the data that led to formation of those regression models. Moreover, it is assumed that no critical effect has been overlooked, which can be tested using power analysis approaches [9]. This ensures that residual variance in the regression models can be attributed to normal analytical- and process variance.
- Optional: Statistical spiking models of each unit operation describing the dependency between varied impurity load and specific impurity clearance (see Section 2.3.2). Identical assumptions as for the regression models must be met.

2.1. Description of Biopharmaceutical Manufacturing Process

This industrial biopharmaceutical process produces a pharmaceutically active recombinant protein and is divided into 7 unit operations. After fermentation using *Escherichia coli* as host cells and recombinant expression of the product, a cell lysis step is performed prior to a precipitation step and clarification. After these primary recovery steps, three preparative chromatographic columns are performed to clear the product from impurities. A final ultrafiltration/diafiltration is performed to adjust the product concentration in drug substance. Two process-related impurities as well as 2 product-related impurities were defined as the major CQAs and herein are modeled within the IPM. Since the analytical quantification of those CQAs was only possible in the load of the first chromatographic step, this step was set as input to the IPM. A summary of the relevant unit operations for modeling, their varied PPs within DoEs, the relative standard deviation of those PPs between large scale (LS) runs, and the monitored CQAs is given in Table 1.

Table 1. Available data sets, process parameters, and monitored critical quality attributes (CQAs) for each unit operation included in the integrated process model (IPM). CC is abbreviation for chromatography column, PCI stands for process-related impurities and PRI product-related impurities.

UO	Available Data Sets	PPs Varied in DoEs	Rel. Std. of PPs between LS [%] [1]	Std/NOR [%] [2]	Monitored CQAs
CC 1	9 manufacturing runs 13 DoE runs with definitive screening design	pH [–]	1.61	46	
		Column loading density [g/L]	12.05	50	
		Wash Strength [mM]	5.00	62	
		Elution strength [mM]	5.00	44	
		End pooling [CV]	1.36	40	
CC 2	9 manufacturing runs 11 DoE runs using full factorial design 1 spiking run with increased PRI 1 concentration in load 1 spiking run with increased PCI 1 concentration in load	pH [–]	0.79	30	PCI 1, PCI 2, PRI 1, PRI 2
		Column loading density [g/L]	4.84	20	
		Gradient slope [% of Buffer]	5.00	-	
CC 3	9 manufacturing runs 9 DoE runs using definitive screening design	pH [–]	0.92	35	
		Column loading density [g/L]	12.78	30	
		Gradient slope [% of Buffer]	5.00	-	
		Wash Strength [mM]	5.00	50	

[1] Relative standard deviation to the set-point of the process parameters; [2] Ratio of one standard deviation to the normal operating range.

2.2. Scope of IPM and Sampling Distribution of PPs

Due to the limited amount of quantitative analytical data of the CQAs before chromatography column 1, the starting distribution of each CQA at the first chromatography step was assumed to be normally distributed with mean and standard deviation estimated from measured CQA distribution of LS runs. From this starting point, the following unit operations chromatography column 1, 2 and 3 were modeled. The pool of chromatography column 3 was regarded as very similar to drug substance since no further clearance formation was expected at the ultrafiltration/diafiltration step.

For the MC workflow, we have to choose a realistic distribution of the large scale variation in process parameters in order to incorporate process-related variability. Results of the MC simulation are dependent on the sampling strategy for the process parameters at each simulation. Often pseudo-random numbers are replaced by quasi-random numbers or Latin hypercube sampling [10,11] for better overview of possible outcomes. However, for realistic risk assessment, we want our sampling to be representative for the process, therefore classical pseudo-random numbers were used for sampling. Existing variance of process parameters has been estimated from current large scale runs as listed in Table 1. We assumed a multivariate normal distribution for all process parameters centered at their mean (target of operation) and variance from large scale runs without any covariance between the process parameters. This is a suitable simplification since process parameters are controlled independently from each other. In general, this is not a prerequisite for the IPM and might be adapted for other processes, where additional information of potential correlation between the process parameters exists.

2.3. Impurity Clearance Models

Since it was aim of the IPM to model the final distribution of each of the four major CQAs (i.e., the specific concentration of each impurity) and the product in the drug substance, their reduction from load of chromatography column 1 until drug substance needs to be described mathematically.

In order to estimate the specific CQA concentration after a unit operation (pool) using the specific load concentration of this CQA, specific clearances (SCs) were used (Equation (1)):

$$\text{Specific Clearance} = \text{SC} = \frac{c_{\text{CQA,load}}}{c_{\text{CQA,pool}}} \tag{1}$$

where $c_{\text{CQA,load}}$ and $c_{\text{CQA,pool}}$ is the specific CQA concentration, defined as the amount of impurity per amount of product, for load and pool, respectively.

For modeling the product a similar approach was chosen using step yields (SY) instead of SC (Equation (2)):

$$\text{Step Yield} = \text{SY} = \frac{p_{\text{pool}}}{p_{\text{load}}} \tag{2}$$

where p_{pool} and p_{load} are the amounts of product in pool and load, respectively.

Two major impacting sources specific clearances have been considered here: (i) Impact of potential critical process parameters, which have been purposefully selected in a risk assessment and (ii) specific amount of impurity load per column volume. Those types of models are described in more detail in the following two Sections 2.3.1 and 2.3.2, respectively. In case it was not possible to find any PPs that significantly impact on the clearance, the mean clearance from LS was taken as a constant model (see Section 2.3.1 for details). We summarize all found models in Table 2.

Table 2. Summary of the presence of models that describe the relationship of a CQA specific clearance factor as a function of PPs (indicated by "DoE") or the impurity loading density of the respective CQA ("Spiking") and the respective *p*-value of the regression. In cases where no significant function of PPs on a CQA clearance could be found, mean large scale clearance was assumed indicated by "LS clearance" in the table. CC is abbreviation for chromatography column, PCI stands for process-related impurities and PRI product-related impurities.

CQA/Unit Operation	CC 1	CC 2	CC 3
PRI 1	DoE (linear, $p = 0.09$)	LS clearance + Spiking ($p = 0.00$)	DoE (quadratic, $p = 0.01$)
PRI 2	DoE (linear, $p = 0.01$)	LS clearance	LS clearance
PCI 1	DoE (quadratic, $p = 0.00$)	LS clearance + Spiking ($p = 0.04$)	DoE (quadratic, $p = 0.00$)
PCI 2	LS clearance + Spiking (linear, $p = 0.00$)	LS clearance	LS clearance + Spiking (linear, $p = 0.00$)
Yield	DoE (linear, $p = 0.00$)	LS clearance	DoE (quadratic, $p = 0.00$)

2.3.1. Clearance and Yield as a Function of Process Parameters (DoE Models)

As a general good practice in PV stage 1, after a purposeful selection of potential impacting process parameters, their impact on the SCs and the SY has been tested within DoEs. For reasons of simplicity, we will only show the modeling approach for SC in the following two sections and not for step yields, since both approaches are identical when exchanging SC with step yield. Experimental designs were chosen (see Table 1 for number of DoE runs and design) and linear models were established according to the form (Equation (3)):

$$\text{SC} = \text{PP} * \beta_{\text{PP}} + \beta_0 + \varepsilon \tag{3}$$

where SC is a ($n \times 1$) vector of the measured specific clearances, PP is a ($n \times p$) matrix of the process parameter settings of each DoE run, β_{PP} are the regression coefficients, and β_0 is the intercept. The process of selecting a subset of significant process parameters was accomplished by means of stepwise regression using multiple linear regression (MLR) package in inCyght (inCyght 2017.03,

Exputec GmbH, Vienna, Austria). In this stepwise procedure, parameters showing a partial p value below 0.05 were allowed to enter the model starting with those parameters having the lowest partial p value. Partial p values of parameters can change as other parameters are included in a multivariate model. Therefore, after each time including a new parameter in the model, it is checked if p values of the existing parameters have increased and those parameters showing an p value larger than 0.1 will be excluded from the model. This including/excluding procedure was applied iteratively to achieve the optimal model, starting with the most significant parameter and was repeated as long as the model structure did not change any more. Thereby, $\hat{\beta}_{PP}$ and $\hat{\beta}_0$ could be estimated. The herein obtained models and their respective statistical quality measures are summarized in Table S1 of the Supplementary materials.

A new prediction for SC ($\widehat{SC}$) for randomly selected set of process parameters of the *i*th MC simulation can be obtained by (Equation (4)):

$$\widehat{SC}(PP^{(i)}) = PP^{(i)} * \hat{\beta}_{PP} + \hat{\beta}_0 \tag{4}$$

The prediction error of the mean model response was assumed to be normally distributed with: $N\left(\widehat{SC}(PP^{(i)}), \sigma^2_{\widehat{SC}(PP^{(i)})}\right)$. Where $\sigma_{\widehat{SC}(PP^{(i)})}$ was calculated using (Equation (5)):

$$\sigma_{\widehat{SC}(PP^{(i)})} = s_{SC} * \sqrt{\frac{1}{n} + h_i} \tag{5}$$

with the leverage of the new data point: $h_i = \text{diag}(PP^{(i)}(PP'PP)^{-1}PP^{(i)'})$, the residual standard error: $s_{SC} = \sqrt{\frac{\sum(SC_i - \widehat{SC}_i)^2}{n-p-1}}$ if p are the number of parameters and n the number of observations. A random sample $\text{rand}(N\left(\widehat{SC}(PP^{(i)}), \sigma^2_{\widehat{SC}(PP^{(i)})}\right))$, using MATLAB (MATLAB, The MathWorks Inc., R2015b, Natick, MA, USA) function randn, was taken from this prediction error distribution for each Monte Carlo simulation *i* and added to the mean prediction, obtaining the specific clearance impacted by PPs for each unit operation (Equation (6)):

$$\widetilde{SC}(PP^{(i)}) = \text{rand}(N\left(\widehat{SC}(PP^{(i)}), \sigma^2_{\widehat{SC}(PP^{(i)})}\right)) \tag{6}$$

For responses where no further spiking models have been available, the specific CQA concentration of the pool of the *u*th unit operation was calculated to (Equation (7)):

$$c_{CQA,pool,u}{}^{(i)} = \frac{c_{CQA,load,u}{}^{(i)}}{\widetilde{SC}(PP^{(i)})} = \frac{c_{CQA,pool,u-1}{}^{(i)}}{\widetilde{SC}(PP^{(i)})} \tag{7}$$

Note that here the concatenation of the unit operations occurs since the specific CQA concentration of the pool of unit operation $u - 1$ is set equal to the load of unit operation u.

If no significant effects of any PP on an impurity clearance of a certain unit operation could be detected, a constant impurity clearance was assumed within the entire design space. This was modeled by the mean clearance of the LS runs and variance of the LS runs. In those cases, for each unit operation the specific clearance of the *i*th MC simulation reduces to (Equation (8)):

$$\widetilde{SC}(PP^{(i)}) = \text{rand}(N\left(\overline{SC}_{LS}, \sigma^2_{SC_{LS}}\right)) \tag{8}$$

where $\overline{SC}_{LS}$ and $\sigma^2_{SC_{LS}}$ is the mean SC and the variance from LS runs, respectively.

2.3.2. Increased Clearance Due to Varied Spiking of Impurities

During process development and design, increased impurity levels were spiked on chromatographic preparative columns in order to show elevated clearance capacity. In more detail, during those spiking studies, it was shown that the impurity clearance increases with increasing impurity loading density (ILD = $\frac{\text{loaded impurity amount}}{\text{column volume}}$) up to the tested level. Additionally, the same relationship of increased impurity clearance at increased impurity loading densities was found for large scale runs, where the impurity loading varies for each run due to variation in fermentation and previous purification unit operations. Since the ILDs were not included within DoE approaches as an independent DoE factor, we followed a two-step approach to incorporate altered clearance at varying ILD.

In the first step, linear regression on SC as a function of ILD was applied to identify significant correlations. Having such a regression model in place, for a specific ILD in the ith MC simulation an estimate for the SC could be obtained ($\widehat{SC}(\text{ILD}^{(i)})$) (Equation (9)):

$$\widetilde{SC}\left(\text{ILD}^{(i)}\right) = \text{rand}\left(\text{N}\left(\widehat{SC}\left(\text{ILD}^{(i)}\right), \sigma^2_{\widehat{SC}(\text{ILD}^{(i)})}\right)\right) \tag{9}$$

where $\widehat{SC}(\text{ILD}^{(i)})$ is the mean predicted SC from the linear regression model at the specific $\text{ILD}^{(i)}$ and $\sigma^2_{\widehat{SC}(\text{ILD}^{(i)})}$ is the variance of the mean prediction, which can be obtained analogous to Equation (5). An example of such a spiking model is shown in Figure 1, where an increased loading density of process-related impurity 2 shows a significant ($p = 7 \times 10^{-8}$) increase in specific clearance of process-related impurity 2. Significant (p-value < 0.05 as well as R^2 (explained variance) $- Q^2$ (from leave one out cross validation) difference < 0.3) spiking models were selected for each response/unit operation and are summarized in Table 2 and Table S2 of the Supplementary Materials.

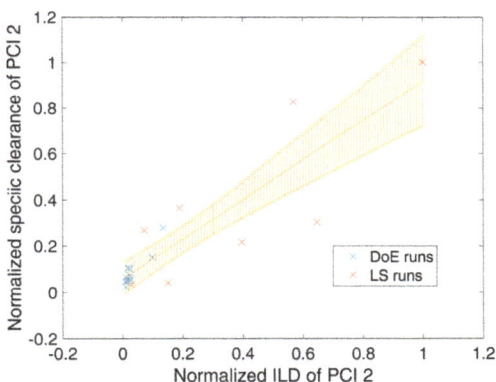

Figure 1. Exemplary plot for dependency of specific clearance (here of process-related impurity 2) against impurity loading density of process-related impurity 2 of DoE runs (blue) and large scale (LS) runs (red). Yellow error bars indicate the mean model prediction error. Normalization has been performed by division of the maximal value for each axis.

Hereafter in the second step, if significant spiking models were available, they were combined with the existing ones as a function of PPs as described in Section 2.3.1. Therefore, for each unit operation, we added the expected clearance increase due to increased ILD to the specific clearance of the ith Monte Carlo simulation impacted by PPs (Equation (10)):

$$\widetilde{SC}^{(i)} = \widetilde{SC}(\text{PP}^{(i)}) * \frac{\widetilde{SC}(\text{ILD}^{(i)})}{\widehat{SC}(\overline{\text{ILD}})} \tag{10}$$

where $\widehat{SC}(\overline{ILD})$ is the SC under mean ILD from DoE runs. The ILD of the simulation i and the unit operation u can be calculated according to (Equation (11)):

$$ILD_u^{(i)} = \frac{c_{load,u}^{(i)} * p_{load,u}^{(i)}}{CV} = \frac{c_{pool,u-1}^{(i)} * p_{pool,u-1}^{(i)}}{CV} \tag{11}$$

where $c_{load,u}^{(i)}$ is the specific concentration of the CQA at the ith simulation and the uth unit operation and $p_{load,u}^{(i)}$ is the product amount modeled by step yield of simulation i and unit operation u, CV is the column volume. Again, the load concentrations and amounts can be expressed by the respective pool concentrations of the previous unit operation ($u - 1$).

Since the impurity loading density was not included within DoE approaches on column steps as an independent DoE factor, we assume that varied impurities do not show interactive effects with other DoE factors (mainly process parameters) within normal operating variance. In order to estimate the risk that the simulation performance is biased by the spiking models and the risk of the above stated assumptions, the IPM was simulated without applying any spiking model. Those results are shown in Figures S1–S4 of the Supplementary materials, where we show that only for product-related impurity 1, process-related impurity 1, and process-related impurity 2, the out of specification chance increases by 0.1%, 0.7%, and 4.2%, respectively. Therefore, the above mentioned assumptions about spiking models and the connection to DoE models can be seen as a minor influence to the overall IPM prediction and valid simplification. Moreover, this can be regarded as a valid simplification since the assumed normal manufacturing variance which is used during IPM simulation of process parameters is well within the normal operating range (NOR, see standard deviation to NOR ratio in Table 1 is often below 30%) and therefore around 99% of the simulated batches are run within NOR. However, we want to note that one could even refine the IPM by including specific impurity concentrations in the load of chromatographic columns as an additional factor in DoE experiments to study that effect in combination with all other DoE factors.

3. Results

3.1. Monte Carlo Approach for Integrated Process Modeling

The main idea behind the integrated process is to concatenate impurity clearance models of each unit operation together to predict the CQA distribution at each intermediate and at drug substance. To account for error propagation during this concatenation, we performed a Monte Carlo approach in four steps:

1. 1000 simulations were performed, each having a different set of PPs ($PP^{(i)}$) for the three modeled unit operations (chromatography column 1/2/3) and different initial specific CQA concentrations ($c^{(i)}_{CQA,init}$) at the load of chromatography column 1, sampled from distributions which were estimated from LS runs. Also the variance in PPs was estimated from LS runs and is indicated by a schematic distribution on the x-axis in Figure 2A,B. Additional increase in simulations did not increase model accuracy and 1000 simulations are a common standard for Monte Carlo simulations [7]. A more detailed description of this step and a list of used process parameters are provided in Section 2.2.

2. For each unit operation, we modeled the specific clearance (SC) of each CQA as a function of the critical PPs and the ILD by multiple linear regression. Each model is associated with a prediction error, which is indicated by the blue shaded area around the found regression line Figure 2A,B. The ILD can be derived from $c_{CQA,load}$ of each unit operation, which equals $c_{CQA,init}$ for the first modeled unit operation and $c_{CQA,pool,u-1}$ for all subsequent modeled unit operations (u).

3. Since $c_{CQA,pool,u}$ can be calculated from SC and $c_{CQA,load,u}$, on the whole, $c_{CQA,pool,u}$ can be seen as a function of PP_u as well as $c_{CQA,init}$ or $c_{CQA,pool,u-1}$, as indicated in the formula of Figure 2A,B, respectively. Thereby the model outputs from multiple unit operations can be stacked together,

which is indicated by black arrows in Figure 2A, more thorough description of which models could be found on which CQA and unit operation is depicted in Section 2.3.

4. Since we performed 1000 simulations, each having different settings in process parameters, we obtained a distribution for the specific CQA concentration in the pool and finally in drug substance, indicated on the y-axis of Figure 2A,B and by the distribution in Figure 2C.

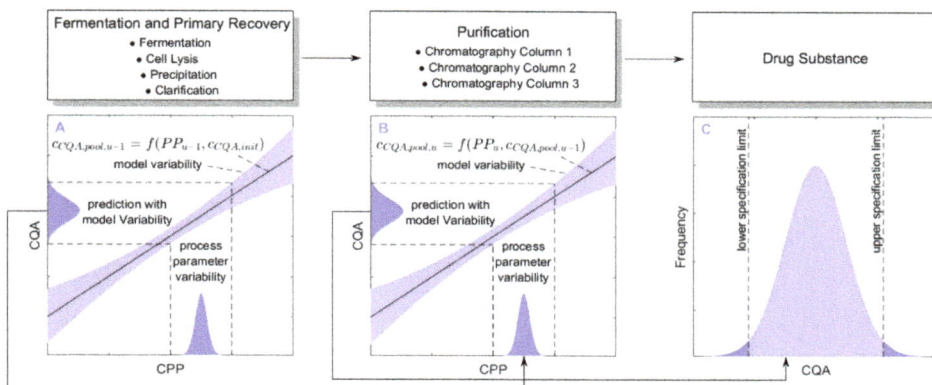

Figure 2. Schematic description of the integrated process model using a Monte Carlo approach: 1000 simulations are performed, each having a different set of process parameters (indicated as distribution on the x-axis of (**A**,**B**)) and initial specific CQA concentration ($c_{CQA,init}$). Multiple linear regression models describe the relationship between the c_{CQA} of the pool of unit operation u (**B**) and the PP of this unit operation as well as the pool concentration of the previous unit operation $u-1$ (**A**). Thereby, models from multiple unit operations (**A**,**B**) are connected to predict the CQA distribution in the drug substance (**C**). Since 1000 simulations are performed, the CQA values form a distribution after each unit operation. The higher the model uncertainty, indicated by blue shaded area around the regression line, the wider the resulting CQA distribution. This ultimately propagates until drug substance, where the chance of out of specification events can be assessed.

3.2. Validation of the IPM Using Observed CQA Distribution in Drug Substance

For model validation, the distribution of the predicted specific CQA concentrations at the pools of each unit operation and drug substance were compared to the measured CQA distribution of LS runs. The OOS chance for the IPM was calculated by simply counting the number of simulations that are above the upper specification limit and dividing by the number of simulations. For the calculation of the OOS chance using the 9 large scale runs, a normal distribution was fitted to the data.

Figures 3–6 show overlays of simulated and observed CQA distribution after each chromatography step for product-related impurity 1 and 2, as well as process-related impurity 1 and 2, respectively. For reasons of data security, all values have been normalized by the maximum observed or simulated CQA value. For the calculation of the observed distributions, all 9 LS runs have been used and have been plotted. CQA distribution after chromatography column 3 (yellow colored bar in Figures 3–6) can be regarded as drug substance since no further purification has been shown to occur at the ultrafiltration/diafiltration step.

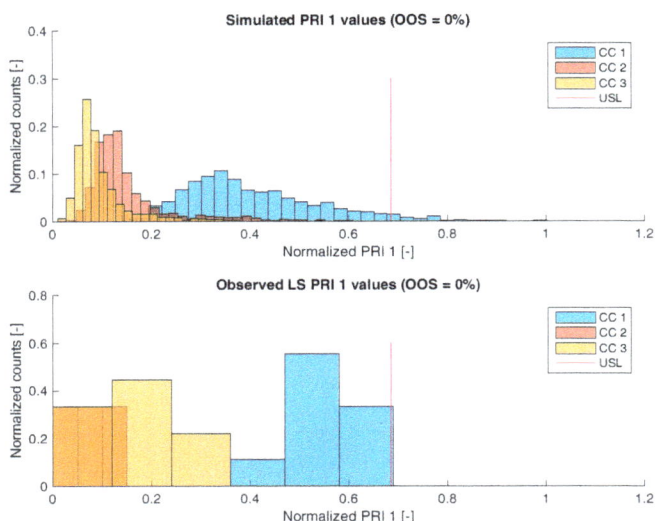

Figure 3. Comparison of simulated (**top**) product-related impurity 1 distribution and observed (**bottom**) product-related impurity 1 from LS after each column step. Normalization was performed by dividing by the maximum observed c_{CQA}.

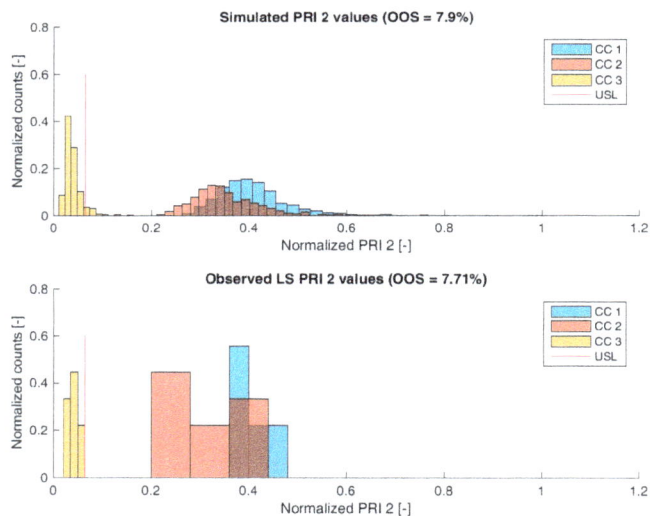

Figure 4. Comparison of simulated (**top**) product-related impurity 2 distribution and observed (**bottom**) product-related impurity 2 from LS after each column step. Normalization was performed by dividing by the maximum observed c_{CQA}.

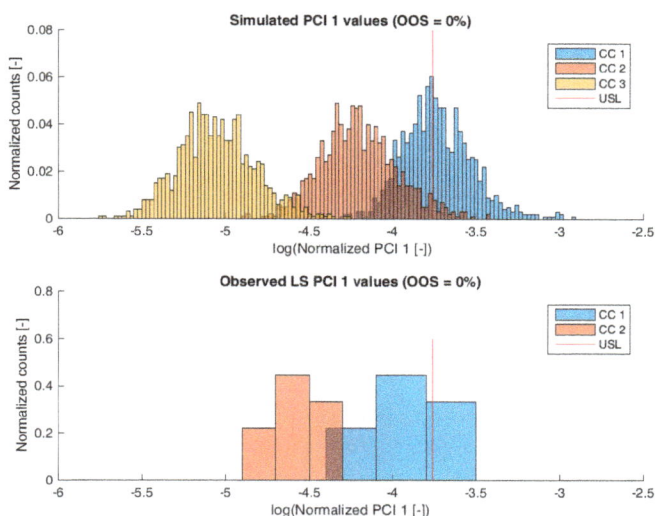

Figure 5. Comparison of simulated (**top**) process-related impurity 1 distribution and observed (**bottom**) process-related impurity 1 from LS after each column step. For chromatography column 3 pool, no process-related impurity 1 value was observed above LoQ, therefore, no histogram bar is plotted for the observed values at chromatography column 3 pool. Normalization was performed by dividing by the maximum observed c_{CQA}.

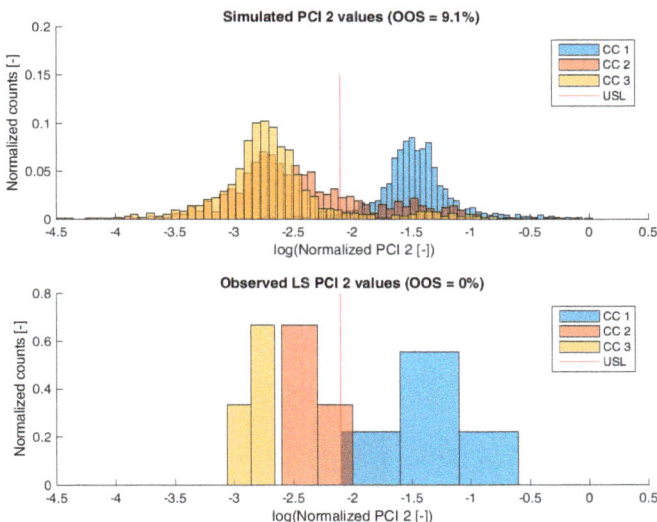

Figure 6. Comparison of simulated (**top**) process-related impurity 2 distribution and observed (**bottom**) process-related impurity 2 from LS after each column step. Normalization was performed by dividing by the maximum observed c_{CQA}.

From visual inspection, the predicted distributions for each CQA nicely overlap with the observed distributions at each chromatography step. This is also reflected in good agreement of simulated and measured OOS probabilities at drug substance level, which are displayed in the title of each subfigure, except for process-related impurity 2. Also, the skewness of the measured CQA distribution is well described by the model (e.g., positive skewness of the product-related impurity 1 distribution at chromatography column 2 in Figure 3). Herein, we regard the model as valid for further investigations such as varying set-point conditions or accelerated variance of PPs.

For process-related impurity 2, the variance of the predicted specific CQA concentrations is larger than the observed variance, especially at chromatography column 3 level, as shown in Figure 6. However, the mean prediction at chromatography column 3 level is very close to the observed runs. The simulated OOS events of the IPM are 9.1% whereas only 0% when calculating from LS data. This gap in predicted versus observed OOS events might be caused by an different mean response of the scale down model at set-point conditions, which was used to conduct the experiments, an overlooked effect of a PP onto this CQA, an overlooked spiking model, or the gap is introduced by the selection of the current large scale runs which show a too low OOS chance. For the first two issues, power analysis for the insignificant models terms needs to identify if additional experiments need to be conducted to make sure that no critical effect has been overlooked [9]. Whereas, the latter possibility indicates a risk that was uncovered by the IPM and has luckily not been observed during LS runs. Herein, counter actions might be taken such as an increase of specific purification capacity in primary recovery.

For product-related impurity 2, the OOS chances for the IPM and the observed data are equally around 7%, as shown in Figure 4. Since for this CQA two statistical models as a function of PPs at chromatography column 1 and chromatography column 3 could be established (Table 2), parameter sensitivity analysis using the IPM can reveal optimization potential to increase process robustness for this CQA.

3.3. Impact of Accelerated Variation in Process Parameters on Drug Substance

Parameter sensitivity analysis (PSA) was performed to assess how a change in set-point or variance of controlled PPs influences OOS events at drug substance. PSA was conducted as follows: Each PP was varied individually regarding its mean and variance and resulting change in OOS events was measured. If interaction effects of parameters have been detected within DoE models, those parameters can be varied simultaneously to study this effect. However, this was not the case for any model established in this study. Moreover, since the model was built only on a segment of all unit operations, we are interested in how an altered performance of the fermentation and primary recovery—leading to an increased impurity burden at the load of chromatography column 1—will impact on drug substance. Therefore, the specific impurity concentration at the loading of chromatography column 1 was also varied in a parameter sensitivity analysis.

Results of an example of such an analysis are shown for product-related impurity 2 (Figure 7), where in panel A the change of OOS events as a function of change in percent of set-point settings of all process parameters is displayed. As can be seen from this subfigure, only a change in pH and wash strength of chromatography column 1 leads to a drastic change in OOS events. This is expected since both factors are part of the DoE model (see Table S1 of Supplementary materials). In more detail, both factors have a favorable direction in terms of reduction of OOS events (lowered pH and increased wash strength). For example, a reduction of the pH value by 10% of the set-point leads to a reduction of OOS events from 7% to around 3%. Interestingly, a change in variance of those two process parameters by ±50% does not impact the OOS events (Figure 7B). This sounds contradictory at first glance, however, since a variance increase to a certain extent will also drive a lot of simulations to the more favorable side (lowered pH and increased wash strength), the overall OOS chance remains similar to the initial estimate. This also emphasizes the well-known fact that optimization should be addressed via a change in the set-point rather than via reduction of variance, which is, in general, even harder to accomplish. A change in initial product-related impurity 2 burden after primary recovery propagates as well into drug substance, which can be explained by the fact that no spiking model could be established for this CQA at any unit operation, as shown in Figure 7C. In detail, a 10% reduction of the specific product-related impurity 2 concentration after primary recovery lowers the OOS events by another 3%. Therefore, it would be favorable to lower the pH of chromatography column 1 and reduce the impurity burden already after primary recovery using prior knowledge or build models that capture the interaction of fermentation and primary recovery parameters on this CQA. Thereby, OOS events could be lowered for product-related impurity 2 down to 1% or less. In order not to increase the OOS probability for another CQA by changing those two process parameters, one would need to also consider their impact on the residual CQAs. This is not shown here since we only wanted to introduce the methodology for a potential application of the IPM and due to reasons of simplicity.

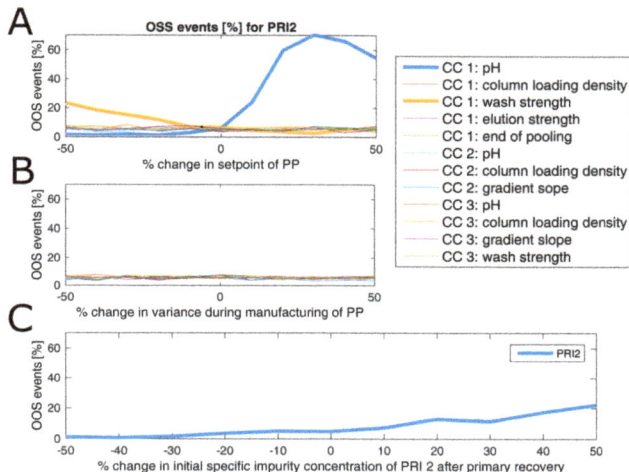

Figure 7. Estimated OOS event for product-related impurity 2 at drug substance as a function of change in set-point (**A**) and variance (**B**) of all PPs as well as a function of increased specific impurity concentration after primary recovery (**C**). Deviations in set-point of pH and salt concentration in wash of chromatography column 1 impact severely on OOS chance, which is not the case when variance in PPs increases by up to 50%. A change of specific product-related impurity 2 concentration at the primary recovery level will also increase OOS chances.

4. Conclusions

Here we have shown how, by using an IPM, it was possible to demonstrate that sufficient process knowledge is available from process development to describe impurity clearance of process-related impurities 1 and 2, as well as product-related impurities 1 and 2. The distributions of simulated and observed CQAs are in good agreement to each other and make it possible to quantify the risk of not meeting product specifications under normal operating conditions, something which is often not possible due to limited large scale runs.

For product-related impurity 1 and process-related impurity 1, both the predicted OOS chance by the IPM as well as the observed OOS chance are numerically close to 0%. Herein, the process design can be validated in respect to those CQAs. In a first application of the IPM within a parameter sensitivity approach, it was possible to identify potential changes in process parameter set-points that will potentially decrease the chance of OOS events for product-related impurity 2 from 7% to 1%. For process-related impurity 2, the mean prediction of clearance within the IPM is similar to that obtained from LS measurements, however, the model predicts a 9.1% chance to be above drug substance specification, whereas current large scale data estimate 0% OOS chance. Since no statistical model could be established that might be used for optimization, process changes might be introduced. Here, IPM can be used within a model life-cycle approach as an enabler in change management. In case parts or entire unit operations are exchanged or included into an existing process design, the IPM can predict the mutual performance of this change in the context of existing clearance capacity. This can be achieved by replacement with statistical models of respective unit operations. Thereby, the overall performance of the changed process design can be assessed in terms of OOS events.

Furthermore, it should be emphasized that this model, in accordance with current opinion, is not finished in the traditional sense, but is expected to incorporate any future experiments and GMP runs for model refinement and application in further PV stages. Thereby, it is expected that new or insufficiently studied dependencies between PPs and CQAs can be incorporated as identified.

Supplementary Materials: The following are available online at www.mdpi.com/2306-5354/4/4/86/s1, Figure S1: Comparison of simulated (**top**) product-related impurity 1 distribution and observed (**bottom**) product-related impurity 1 from LS after each column step, Figure S2: Comparison of simulated (**top**) product-related impurity 2 distribution and observed (**bottom**) product-related impurity 2 from LS after each column step, Figure S3: Comparison of simulated (**top**) process-related impurity 2 distribution and observed (**bottom**) process-related impurity 2 from LS after each column step, Figure S4: Comparison of simulated (**top**) process-related impurity 1 distribution and observed (**bottom**) process-related impurity 1 from LS after each column step, Table S1: Overview of found models based on DoE data, Table S2: Overview of models showing a correlation between specific CQA clearances and CQA load density.

Author Contributions: Thomas Zahel designed the IPM and the Monte Carlo Simulation and wrote of the manuscript. Stefan Hauer assisted in the implementation of the IPM and the Monte Carlo Simulation as well as in designing the presented figures. Eric M. Mueller, Pat Murphy, Sandra Abad, Cécile Brocard, Daniela Reinisch, Patrick Sagmeister and Christoph Herwig assisted in writing the manuscript. Sandra Abad, Elena Vasilieva and Daniel Maurer conducted the necessary experiments in DoE approaches at small scale.

Conflicts of Interest: The authors declare no conflict of interest.

References

1. *Process Validation: General Principles and Practices*; U.S. Department of Health and Human Services: Washington, DC, USA, 2011.
2. Guideline, I.H.T. Pharmaceutical Development Q8 (R2). *Curr. Step* **2009**, *4*, 1–24.
3. Guideline, I.H.T. Quality risk management, Q9. *Curr. Step* **2005**, *4*, 408.
4. Katz, P.; Campbell, C. FDA 2011 process validation guidance: Process validation revisited. *J. GXP Compliance* **2012**, *16*, 18.
5. Peterson, J.J.; Lief, K. The ICH Q8 definition of design space: A comparison of the overlapping means and the bayesian predictive approaches. *Stat. Biopharm. Res.* **2010**, *2*, 249–259. [CrossRef]
6. Herwig, C.; Wölbeling, C.; Zimmer, T. A holistic approach to production control. *Pharm. Eng.* **2017**, *37*, 44–46.
7. Bonate, P.L. A brief introduction to Monte Carlo simulation. *Clin. Pharmacokinet.* **2001**, *40*, 15–22. [CrossRef] [PubMed]
8. Goudar, C.T.; Biener, R.; Konstantinov, K.B.; Piret, J.M. Error propagation from prime variables into specific rates and metabolic fluxes for mammalian cells in perfusion culture. *Biotechnol. Prog.* **2009**, *25*, 986–998. [CrossRef] [PubMed]
9. Iman, R.L. Latin hypercube sampling. *Encycl. Quant. Risk Anal. Assess.* **2008**. [CrossRef]
10. Singhee, A.; Rutenbar, R.A. Why quasi-monte carlo is better than monte carlo or latin hypercube sampling for statistical circuit analysis. *IEEE Trans. Comput. Aided Des. Integr. Circuits Syst.* **2010**, *29*, 1763–1776. [CrossRef]
11. Zahel, T.; Marschall, L.; Abad, S.; Vasilieva, E.; Maurer, D.; Mueller, E.M.; Murphy, P.; Natschläger, T.; Brocard, C.; Reinisch, D.; et al. Workflow for Criticality Assessment Applied in Biopharmaceutical Process Validation Stage 1. *Bioengineering* **2017**, *4*, 85. [CrossRef] [PubMed]

MDPI AG

St. Alban-Anlage 66

4052 Basel, Switzerland

Tel. +41 61 683 77 34

Fax +41 61 302 89 18

http://www.mdpi.com

Bioengineering Editorial Office

E-mail: bioengineering@mdpi.com

http://www.mdpi.com/journal/bioengineering

www.ingramcontent.com/pod-product-compliance
Lightning Source LLC
Chambersburg PA
CBHW051907210326
41597CB00033B/6063